D1581145

GCSE 9-1

FOUNDATION

MATHEMATICS

AQA REVISION & EXAM PRACTICE

Gwenllian Burns,
Catherine Murphy and
Naomi Norman

Authors Gwenllian Burns, Catherine Murphy and Naomi Norman
Editorial team Haremi Ltd
Series designers emc design ltd
Typesetting York Publishing Solutions Pvt. Ltd.
Illustrations York Publishing Solutions Pvt. Ltd.
App development Hannah Barnett, Phil Crothers and Haremi Ltd

Designed using Adobe InDesign
Published by Scholastic Education, an imprint of Scholastic Ltd, Book End, Range Road, Witney, Oxfordshire, OX29 0YD
Registered office: Westfield Road, Southam, Warwickshire CV47 0RA
www.scholastic.co.uk

Printed by Bell & Bain Ltd, Glasgow
© 2017 Scholastic Ltd
1 2 3 4 5 6 7 8 9   7 8 9 0 1 2 3 4 5 6

British Library Cataloguing-in-Publication Data
A catalogue record for this book is available from the British Library.
ISBN 978-1407-16907-1

Acknowledgements
Map on page 101: © Petersfield Town Council (contains Ordnance Survey data © Crown copyright and database right 2010)

Every effort has been made to trace copyright holders for the works reproduced in this book, and the publishers apologise for any inadvertent omissions.

Notes from the publisher
Please use this product in conjunction with the official specification and sample assessment materials. Ask your teacher if you are unsure where to find them.

The marks and star ratings have been suggested by our subject experts, but they are to be used as a guide only.

Answer space has been provided, but you may need to use additional paper for your workings.

How to use this book

Inside this book you'll find everything you need to help you succeed in the new GCSE 9–1 AQA Foundation Mathematics specification. It combines revision and exam practice in one handy solution. Broken down into topics and subtopics, it presents the information in a manageable format. Work through the revision material first or dip into the exam practice as you complete a topic. This book gives you the flexibility to revise your way!

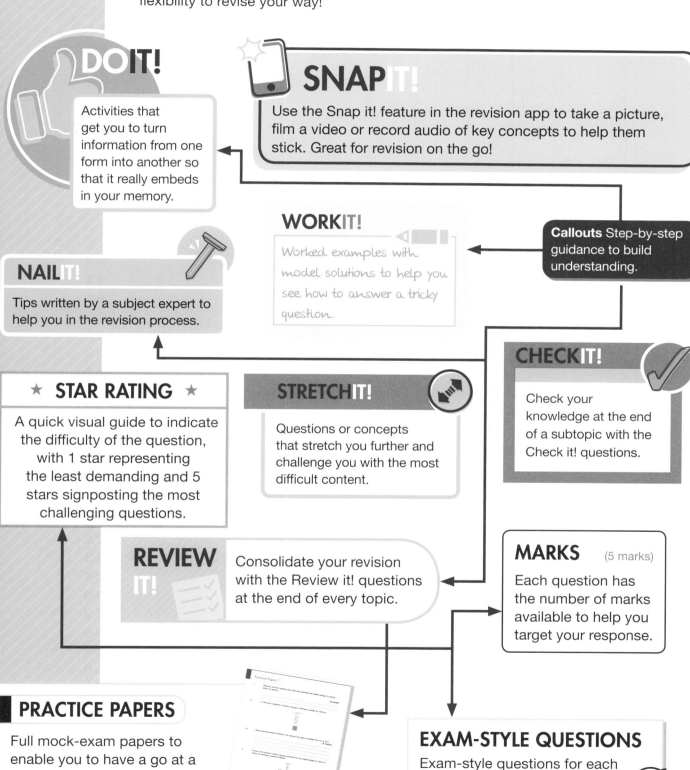

DOIT!

Activities that get you to turn information from one form into another so that it really embeds in your memory.

SNAPIT!

Use the Snap it! feature in the revision app to take a picture, film a video or record audio of key concepts to help them stick. Great for revision on the go!

WORKIT!

Worked examples with model solutions to help you see how to answer a tricky question.

Callouts Step-by-step guidance to build understanding.

NAILIT!

Tips written by a subject expert to help you in the revision process.

CHECKIT!

Check your knowledge at the end of a subtopic with the Check it! questions.

★ STAR RATING ★

A quick visual guide to indicate the difficulty of the question, with 1 star representing the least demanding and 5 stars signposting the most challenging questions.

STRETCHIT!

Questions or concepts that stretch you further and challenge you with the most difficult content.

REVIEW IT!

Consolidate your revision with the Review it! questions at the end of every topic.

MARKS   (5 marks)

Each question has the number of marks available to help you target your response.

PRACTICE PAPERS

Full mock-exam papers to enable you to have a go at a complete paper before you sit the real thing!

EXAM-STYLE QUESTIONS

Exam-style questions for each subtopic ramped in difficulty.

Revision contents

Topic 4

GEOMETRY AND MEASURES

Topic 5

PROBABILITY

Topic 6

STATISTICS

Exam Practice contents

Topic 4

GEOMETRY AND MEASURES

Topic 5

PROBABILITY

Topic 6

STATISTICS

HOW TO REVISE!

PLAN YOUR REVISION

Get ahead by planning your revision!

Work out the **time** you have available for revising.

Think about when you work at your best. Are you a morning or an evening person?

Allocate **MORE TIME** for the topics you struggle with.

Revision works best in **SMALL BURSTS**, so keep sessions **SHORT AND SWEET**!

Remember to allow time to **PRACTISE** applying what you have revised.

Use your **revision app** to put together a revision timetable.

LOOK AFTER YOURSELF

Help your brain by looking after your whole body!

Take regular **breaks** from revising – your brain needs time to digest information in order to retain it.

HOTEL

Keep **hydrated** by drinking plenty of water – dehydration stops your brain from working at its full capacity.

Regular **exercise** helps stimulate the brain and will help you relax.

Get plenty of **sleep**, especially the night before an exam.

EAT WELL and limit unhealthy snacks – your brain needs fuel for memory and concentration.

Find methods of **relaxation** that work for you throughout the revision period.

BE PREPARED!

Limit potential stress on the day of an exam by getting everything you need ready the night before.

30

Highlight key pieces of written information and use different colours to classify it.

DO IT!

Take an active approach to revision. Create **revision cards**, **lists**, **mind maps**, **flow charts** and **infographics** like this one.

Annotate and summarise your class notes and revision pages.

Revise in pairs or small groups and deliver presentations on topics to each other.

SNAP IT!

Use your phone to take pictures of your revision material so that you can **revise on the go!**

REMEMBERING INFORMATION

Use different methods to digest information!

Create a safe place to **store** all of your revision notes and keep everything together.

Use **mnemonics** or songs to help learn tricky bits of information.

RETRIEVING INFORMATION

Use the information you have remembered!

REVIEW IT!

Revisit the content you have already revised to confirm your understanding.

Practise applying your knowledge using exam-style questions.

READ THE QUESTION CAREFULLY!

Make sure you understand what you are being asked to do.

CHECK IT!

Make sure you know the material using simple **recall** questions.

Mark your work to help you see where you can **improve**.

FOR HIGH-MARK QUESTIONS, SPEND TIME **PLANNING** YOUR ANSWER!

Stick to the **TIME LIMITS** you will need to in the exam.

FINISH

Number

Factors, multiples and primes

NAIL IT!

A divisor is a whole number that will divide into another number.

For example, 2 is a divisor of 8.

NAIL IT!

To find all the factors of a number, list all the pairs that multiply to give that number. Start with 1 and work up!

For example, factors of 20:

1 × 20

2 × 10

4 × 5

SNAP IT! Factors, multiples and primes

Factors

Factors of a number divide into it exactly. For example, the factors of 12 are 1, 2, 3, 4, 6 and 12. Notice that you include 1 and the number itself.

Multiples

To find the multiples of a number, multiply it by 1, 2, 3 and so on. For example, multiples of 5 are 5, 10, 15, 20, 25 and so on.

Prime numbers

A prime number is one with **exactly** two factors. For example, 7 is prime as its only factors are 1 and 7.

Highest common factor

The highest common factor (HCF) of two numbers is the largest number that is a factor of both (divides into both exactly).

Lowest common multiple

The lowest common multiple (LCM) of two numbers is the smallest number that is a multiple of both (they will both divide into it exactly).

WORKIT!

Find the highest common factor of 20 and 36.

The factors of 20 are: 1, 2, ④, 5, 10, 20

The factors of 36 are: 1, 2, 3, ④, 6, 9, 12, 18, 36

The highest factor that occurs in both lists is 4.

The highest common factor of 20 and 36 is 4.

WORKIT!

Find the lowest common multiple of 10 and 15.

Multiples of 10 are: 10, 20, ㉚, 40, 50, 60, 70, 80...

Multiples of 15 are: 15, ㉚, 45, 60, 75...

The lowest value that appears in both lists is 30.

So the lowest common multiple is 30.

NAIL IT!

To find the lowest common multiple, list as many multiples as necessary until you find one that appears in both lists. Sometimes it is simply the result of multiplying the two numbers together.

For example, the lowest common multiple of 5 and 7 is 5 × 7 = 35

Prime factor decomposition

All numbers can be written as a product of their prime factors. This is called prime factor decomposition.

WORKIT!

Write 180 as a product of prime factors. Write your answer in index form (using indices).

Here is the factor tree for 180.

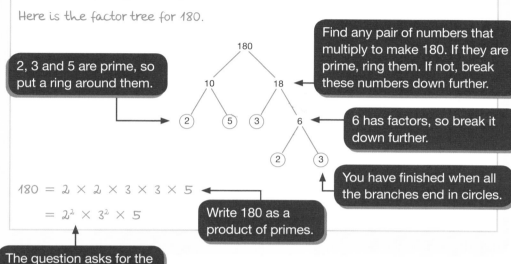

2, 3 and 5 are prime, so put a ring around them.

Find any pair of numbers that multiply to make 180. If they are prime, ring them. If not, break these numbers down further.

6 has factors, so break it down further.

You have finished when all the branches end in circles.

$180 = 2 \times 2 \times 3 \times 3 \times 5$

$\qquad = 2^2 \times 3^2 \times 5$

Write 180 as a product of primes.

The question asks for the answer in index form.

DOIT!

Choose two different two-digit numbers between 10 and 30. Find the highest common factor and the lowest common multiple of the two numbers.

NAILIT!

Never include a 1 in the factor tree, since 1 is not a prime number.

You can use prime factor decomposition to find the highest common factor and the lowest common multiple of two numbers.

To find the highest common factor and lowest common multiple of 30 and 40, first write both numbers as a product of primes, using prime factor decomposition.

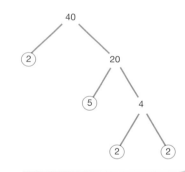

Display the information in a Venn diagram, making sure that any numbers that appear in both trees are written in the intersection (the leaf-shaped area where the circles overlap).

The highest common factor is found by multiplying together any factors that appear in **both** trees. These are the ones in the intersection of the Venn diagram.

$HCF = 2 \times 5 = 10$

The lowest common multiple is found by multiplying all the factors in the Venn diagram:

$LCM = 3 \times 2 \times 5 \times 2 \times 2 = 120$

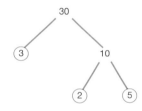

See page 181 for more explanation of set notation.

WORKIT!

Given the following facts, find the highest common factor and lowest common multiple of 18, 20 and 30.

$18 = 2 \times 3 \times 3$

$20 = 2 \times 2 \times 5$

$30 = 2 \times 3 \times 5$

Highest common factor = 2

Lowest common multiple = 2 × 2 × 3 × 3 × 5 = 180

> Draw a Venn diagram to display the information.

> The only value that appears in the intersection of all three circles is 2.

> Multiply together all the values that appear in the Venn diagram.

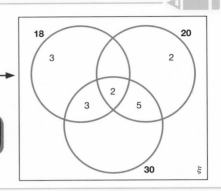

STRETCHIT!

Prime factors can be small but they may also be very large. Although you won't be asked to find large prime factors in an exam, they are very useful for encrypting and decoding digital data. When a number only has huge prime factors they are extremely difficult to find!

CHECKIT!

1 Here is a list of five numbers:

1   5   12   30   45

Which numbers in the list are

a   prime

b   factors of 12

c   **not** multiples of 2

d   divisors of 10?

2 Find the highest common factor and lowest common multiple of 150 and 70.

3 Write 90 as a product of prime factors using index form.

> Index form means numbers with powers.

> The product is what you get when you multiply things together.

4 Given that

$120 = 2 \times 2 \times 2 \times 3 \times 5$

$70 = 2 \times 5 \times 7$

$30 = 2 \times 3 \times 5$

a   find the highest common factor

b   find the lowest common multiple.

5 The highest common factor of a pair of numbers is 6. Both numbers are between 10 and 20. What are they?

Ordering integers and decimals

Ordering positive and negative integers

A negative number is smaller than zero.

As you move left along the number line, the numbers get lower.

−10 is lower than −4.

NAILIT!

If the temperature is −10°C it will be colder than a temperature of −1°C, so −10 is lower than −1.

Place value

In a number, for example 3108475.926, the value of each digit depends on its position in the number − its **place value**.

millions	hundreds of thousands	tens of thousands	thousands	hundreds	tens	units	tenths	hundredths	thousandths
3	1	0	8	4	7	5	9	2	6

The 3 is worth 3 million: 3 000 000

The 7 is worth 7 tens: 70

The 2 is worth 2 hundredths: 0.02

> The units are the numbers from 1 to 9. You may also know them as the 'ones'.

Ordering decimals

To write decimals in order of size, compare the whole number part, then tenths, then hundredths and so on. Work from left to right.

WORKIT!   3.4 3.900

a Which is larger, 3.431 or 3.9?

3.9 is larger than 3.431

> First look at the whole number parts:
> **3**.431 or **3**.9   they are the same
> Next compare the tenths:
> 3.**4**31 or 3.**9**   9 is greater than 4

b Put these numbers in ascending order.

−3, 3.03, 0.3, −0.03

−3, −0.03, 0.3, 3.03

−3, −0.03, 0.3, 3.03

> Think about the position of each number on the number line.

DOIT!

Write down a large number using all the digits 1−9. Write down the value of each digit.

NAILIT!

Ascending means going up.

Descending means going down.

SNAPIT!   Comparing numbers

To compare numbers, use the symbols:

Symbol	Meaning	Symbol	Meaning
=	equal to	≠	not equal to
<	less than	>	greater than
≤	less than or equal to	≥	greater than or equal to

NAILIT!

The wide end (or 'mouth') of the inequality sign always points towards the larger value.

For example, 15 > 2 means '15 is greater than 2'.

=	≠
<	>
≤	≥

✓ CHECKIT!

1 Decide whether the following statements are true or false.

   a −4 < −7   **d** 3.41 ≥ 3.14

   b 4.1 > 4.01   **e** 0.909 ≤ 0.99

   c 0.1 ≠ 0.01

2 Write these numbers in **descending** order:

   −2.5, −4.2, −0.3, −1.5, −7.2

3 Write these numbers in **ascending** order:

   0.412, 0.124, 0.442, 0.049, 1.002

4 Replace each ? symbol with > or <.

   a 0.25 ? 0.52

   b 1.07 ? 1.1

   c 13.4 ? 13.099

Calculating with negative numbers

Finding differences

A number line is useful to find the difference between a positive and negative number.

The difference between -2 and 4 can be calculated by considering the 'jumps' that are taken along the number line to move from -2 to 4.

The difference between -2 and 4 is $2 + 4 = 6$.

First jump from -2 to 0, then jump from 0 to 4.

Adding and subtracting negative numbers

SNAPIT!   **Adding and subtracting negative numbers**

> Adding a negative number is the same as subtracting a positive number. Subtracting a negative number is the same as adding a positive number.

WORKIT!

Work out

a $-4 + -2$

$-4 - 2 = -6$ ◀ Adding a negative number is equivalent to subtracting, so you can rewrite the calculation as a subtraction.

b $3 - -5$

$3 + 5 = 8$ ◀ Subtracting a negative number is equivalent to adding, so you can rewrite the calculation as an addition.

Multiplying and dividing negative numbers

SNAPIT!   **Multiplying and dividing negative numbers**

> When multiplying and dividing two negative numbers, remember the rules:
> - If the signs are the same the answer will be positive.
> - If the signs are different the answer will be negative.

DO IT!

Make a mind map about negative numbers. Include everything you know about working with negative numbers.

STRETCH IT!

If you multiply together three negative numbers, will the solution be positive or negative? Is this always true?

WORK IT!

Work out

a -2×8

$-2 \times 8 = -16$ ←

> Multiply: $2 \times 8 = 16$
> The signs are different so the answer will be negative.

b $-20 \div -5$

$-20 \div -5 = 4$ ←

> Divide: $20 \div 5 = 4$
> The signs are the same, so the answer is positive.

c $-3 \times -5 \times -1$

$-3 \times -5 \times -1 = 15 \times -1$

$= -15$

> Work out -3×-5 first.

CHECK IT!

1 Work out

a $-8 + -3$

b -9×-11

c $18 \div -3$

d $14 - -4$

e $-8 + 8$

f $12 - -15 - 2$

2 Two of the numbers below are multiplied together.

Which pair of numbers gives the lowest answer?

| -5 | -8 | 4 | 9 |

3 The average daily temperature in Moscow in December is $-8°C$. At the North Pole the average daily temperature is $-40°C$.

How much colder is it at the North Pole than in Moscow?

Multiplication and division

Division

You can use short division to divide a large number by a single-digit number. The number you are dividing by is called the divisor. You work from left to right, dividing into each digit in turn, as shown in the Work it! example.

WORKIT!

Work out 348 ÷ 3.

Write the number to be divided within the lines and the divisor outside the lines.

Work left to right, dividing each digit by 3.

If 3 doesn't divide exactly into the number, work out how many times it will go completely and write this above the line. 'Carry' any remainder on to the next number.

The 1 that has been carried on to the 8 is worth 10 in the units column, so divide 18 by 3.

Sometimes the divisor will not divide exactly. There are two ways to find the exact solution.

Method 1: Write the solution as a mixed number

$$5\overline{)1\ ^12\ ^24}\quad \text{remainder 4}$$

The remainder still needs to be divided by 5, so write:

$$124 \div 5 = 24\tfrac{4}{5}$$

Write the remainder as the numerator of the fraction and the divisor as the denominator.

Method 2: Work out the solution as a decimal

$$5\overline{)1\ ^12\ ^24\ ^40}$$

Insert a decimal point and write as many zeros as you like after the number. Keep working as before. When the divisor divides exactly you can stop adding more zeros.

$$124 \div 5 = 24.8$$

NAILIT!

$124 = 124.0$
$= 124.00 = 124.000$

Long division

You can use long division to divide a large number by a two-digit or larger number. As with short division, you work from left to right. This time you take two or more digits at once (because your divisor is bigger than 9). See the Work it! on the next page for how to do this step by step.

WORKIT!

Work out 746 ÷ 15.

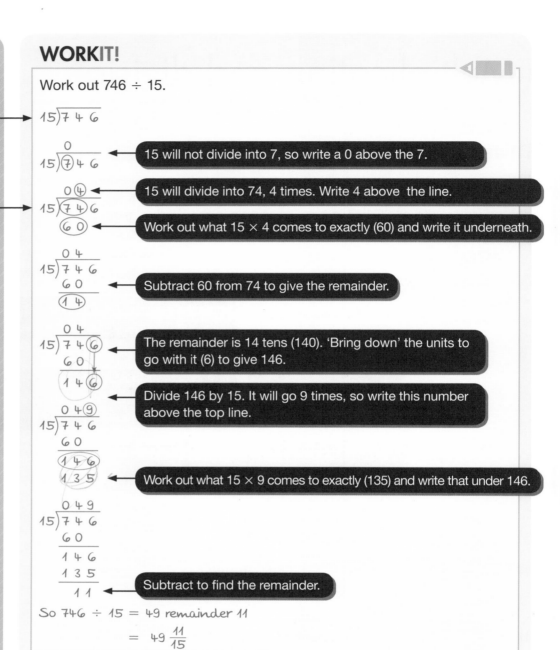

Write the number to be divided within the lines and the number to divide by (divisor) outside the lines.

15 will not divide into 7, so write a 0 above the 7.

Now look at the first pair of digits. Will 15 divide into 74?

15 will divide into 74, 4 times. Write 4 above the line.

Work out what 15 × 4 comes to exactly (60) and write it underneath.

Subtract 60 from 74 to give the remainder.

The remainder is 14 tens (140). 'Bring down' the units to go with it (6) to give 146.

Divide 146 by 15. It will go 9 times, so write this number above the top line.

Work out what 15 × 9 comes to exactly (135) and write that under 146.

Subtract to find the remainder.

So 746 ÷ 15 = 49 remainder 11

$$= 49\frac{11}{15}$$

Roll a dice three times to produce a 3-digit number. Roll it twice more to give a 2-digit number.

Divide the 3-digit number by the 2-digit number and check your answer on a calculator.

Multiplication

Use short multiplication to multiply a one-digit number by a number with two or more digits.

WORKIT!

Work out 172 × 8.

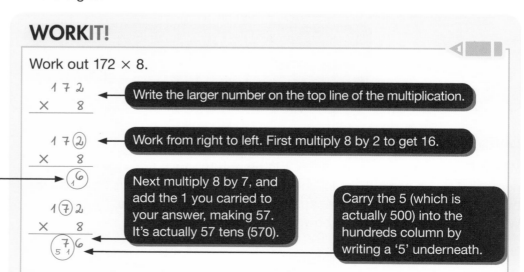

Write the larger number on the top line of the multiplication.

Work from right to left. First multiply 8 by 2 to get 16.

Write the 6 of 16 under the answer line. Carry the 10 into the tens column by writing a '1' underneath.

Next multiply 8 by 7, and add the 1 you carried to your answer, making 57. It's actually 57 tens (570).

Carry the 5 (which is actually 500) into the hundreds column by writing a '5' underneath.

$$\begin{array}{r} ①7\ 2 \\ \times\quad 8 \\ \hline ①③7\ 6 \\ {}_5\ {}_1 \end{array}$$

Next multiply 8 by 1 – don't forget to add the 5 that you carried. This makes 13 (hundreds). Write the 1 in the thousands column as you are not doing any more multiplication.

Long multiplication

Use long multiplication to multiply a two-digit (or more) number by a two-digit number.

WORKIT!

Work out 126 × 43.

Set out the multiplication like this. Break down the number you are multiplying by into units and tens.

First multiply each digit by 3. Don't forget to include any values you carry.

Next multiply each digit by 40: put a zero in the units column and then multiply by 4.

Complete the calculation.

Add together the two values.

126 × 43 = 5418

DOIT!

To check that 348 ÷ 3 = 116, which calculation should you do?

116 × 3 or 116 ÷ 3

Inverse operations

To check the answer to a calculation you can use the inverse operation. An inverse operation 'undoes' what the operation does.

CHECKIT!

Do not use a calculator for these questions.

1 Work out
  a 235 × 9
  b 924 × 61

2 Work out
  a 235 ÷ 5
  c 2146 ÷ 17
  b 4128 ÷ 8

Use inverse operations to check answers to 1 and 2.

3 In a factory, 265 pencils are packed into boxes of 8.
  a How many boxes are filled?
  b How many pencils are left over?

4 Work out £365 ÷ 4.

5 A cinema charges £9 per ticket. If a class of 32 students goes to the cinema, what is the total cost of tickets?

6 Calculate 923 ÷ 3 and give your answer as a mixed number.

7 Work out 823 × 35.

8 Eggs are packed into boxes of 12. How many complete boxes can be filled with 450 eggs?

9 Aaron is calculating 726 × 52. Explain what he has done wrong.

$$\begin{array}{r} 7\ 2\ 6 \\ \times\ 5\ 2 \\ \hline 1\ 4\ 5\ 2 \times 2 \\ 3\ 6③0 \times 50 \\ \hline 5\ 0\ 8\ 2 \end{array}$$

Calculating with decimals

To add or subtract decimals, line up the decimal points.

WORKIT!

Aisha has a tank that holds 50 litres of petrol. At a petrol station she adds 37.5 litres. If the tank already contains 9.02 litres, how much space is left in the tank?

```
          4 9 9¹
    3 7 . 5 0      5̶0̶.0̶0̶
  + 0 9 . 0 2    − 4 6.5 2
  ─────────      ─────────
    4 6 . 5 2      0 3.4 8
```

There are 3.48 litres of space available.

NAILIT!

When adding and subtracting decimals, you can fill in any spaces with a '0' – this will help you to avoid making place value mistakes.

To multiply decimals, first do the calculation without any decimal point, then put a decimal point into the correct place in the answer.

Rewrite the questions, removing the decimal points.

In total, there are three digits after the decimal point in the question, so there must be three in the answer.

WORKIT!

Work out 0.12×0.7

12×07

$12 \times 7 = 84$ ← Multiply.

$0.12 \times 0.7 = 0.084$

Divide decimals in exactly the same way as you divide whole numbers – but remember to keep the decimal point in place!

WORKIT!

Calculate $47.52 \div 3$

NAILIT!

Always check your answers to see if they are sensible.

$15 \times 3 = 45$

Therefore $47.52 \div 3$ must be approximately 15.

✓ CHECK IT!

Do not use a calculator for these questions.

1 Work out

a $3.4 - 1.07$

b $19.3 + 5.091$

c 0.05×0.7

d 3.21×1.9

e $7.815 \div 5$

2 Mohammad buys 24 first class stamps costing £0.64 each. How much change does he get from a £20 note?

3 In a restaurant Erica and Freya split the bill. Erica pays twice as much as Freya. If the total bill was £82.38 how much do they each pay?

Rounding and estimation

Rounding decimals

You can round to a number of **decimal places** – this gives the number of digits after the decimal point.

You can also round to a number of **significant figures** – a significant figure is any digit after the first **non-zero** digit.

In the number 3879 the digit 3 is the first significant figure, the 8 is the second significant figure, the 7 is the third significant figure and the 9 is the fourth significant figure.

In the number 0.000135 the first four zeros are **not** significant figures. The digit 1 is the first significant figure, the 3 is the second significant figure and the 5 is the third significant figure.

NAIL IT!

Round the digit up if the following digit is 5 or more.

Leave the digit as it is if the following digit is 4 or less.

NAIL IT!

Zeros **do** count as significant figures if they are **after** the first significant figure.

For example, in the number 3072 there are four significant figures: 3, 0, 7 and 2.

WORK IT!

Round 3.45901 to 2 decimal places.

3.46 ◄— The last digit is rounded up when the number after it is 5 or more.

WORK IT!

1   Round 4729 to 2 significant figures.

4700 ◄— The second digit is not rounded up since the number after it is less than 5. All the other digits are replaced with zeros.

2   Round 0.00462 to 1 significant figure.

0.005 ◄— The 4 is rounded up since the number after it is 5 or more.

Rounding can be used to check answers in calculations, particularly when you are using a calculator as it is easy to make mistakes.

WORK IT!

Jayshuk enters 0.8 × 4.5 into his calculator and gets the answer 36.

Explain how he **knows** he has entered the calculation incorrectly.

0.8 is slightly less than 1, so if you multiply 0.8 by 4.5 you should get an answer smaller than 4.5.

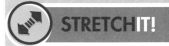

STRETCH IT!

Round 0.99999 to
a 1 decimal place
b 2 decimal places
c 3 decimal places.
What do you notice?

DO IT!

You can use a spreadsheet or an online tool to round numbers for you.

Have a go at rounding the number 3.48720912 to different numbers of significant figures or decimal places and check your answers using a spreadsheet or an online tool.

Error intervals

An **error interval** is the range of values that a rounded number could be.

The **upper bound** is the upper limit for the largest value it could be.

The **lower bound** is the smallest value it could be.

To write an error interval, work out the upper and lower bounds.

For example, if $x = 12.3$ to 1 decimal place you can use inequality notation to write the error interval.

The smallest value that x could be is 12.25.

Use the $\leq$ sign, since 12.25 would round to 12.3.

$$12.25 \leq x < 12.35$$

The largest value that x could be is very close to 12.35.
Use the $<$ sign, since 12.35 would not round to 12.3, but anything smaller would. x could be 12.34999 for example.

WORKIT!

The width of a washing machine is given as 55 cm to the nearest cm. What is the error interval for the width of the machine?

$54.5 \leq \text{width} < 55.5$

The smallest value that would round to 55 is 54.5, so this is the lower bound.

Any number that is 55.5 or larger will round up to 56 not down to 55 so this is the upper bound.

Estimation

Estimation means finding an **approximate** answer. You can find approximate answers by rounding.

NAILIT!

Show **all** your workings — it is good practice to show all the rounded values you are using in your calculation. It also means you can check your working later.

NAILIT!

Always check your answers to see if they are sensible! Is £50 a sensible price to pay for petrol?

WORKIT!

Petrol costs 109p per litre. Angela buys 54 litres of petrol.
Work out the approximate total cost.

$100 \times 50 = 5000\text{p}$

$5000\text{p} = £50$

Round each of the values to 1 signficant figure – this makes the calculation easy!

Don't forget to give the answer in a way that makes sense – you'd never give a price as 5000p!

STRETCHIT!

Mr Appleyard measures a rectangular room to calculate the area of carpet needed. To the nearest metre, he measures the room as 6 m by 8 m.

Work out the area of carpet he should buy to ensure he has enough to cover the whole room. Is your answer an underestimate or overestimate?

CHECKIT!

1 Round each number to the level of accuracy given.

   a 0.34901   2 decimal places

   b 12.092   1 significant figure

   c 32.609   1 decimal place

   d 33 098   3 significant figures

2 The following numbers are rounded to the degree of accuracy shown. Write down the error interval for each number. Use x to represent the number in each case.

   a 200 rounded to the nearest hundred

   b 6 rounded to the nearest whole number

   c 3.2 rounded to 1 decimal place

   d 5.06 rounded to 3 significant figures

3 Round each number to 1 significant figure to find an approximate answer to:

$$\frac{31.45902}{0.491 \times 6.498}$$

4 Without working out the correct answers, which of these do you **know** is false? Explain your answer.

   a $3.5 \times 7 = 24.5$

   b $18 \times 0.9 = 1.62$

   c $19 \div 0.25 = 7.6$

5 An electricity company has two different packages:

Night-time low

1.622 p per unit from 8 pm to 6 am

2.315 p per unit from 6 am to 8 pm

One tariff

1.923 p per unit all day

Tarik uses 2.32 units between 8 pm and 6 am and 7.151 units between 6 am and 8 pm.

Use this information to work out which package Tarik should choose.

Electricity charges

Delivery tariff

Night-time low	1.622 p (8 pm – 6 am) 2.315 p (6 am – 8 pm)
One tariff	1.923 p
Total charges	

Converting between fractions, decimals and percentages

Percentages to fractions

'Per cent' means 'out of 100', so any percentage can be written as a fraction by writing it over 100, then cancelling the fraction down.

$$45\% = \frac{45}{100} = \frac{9}{20}$$

Percentages to decimals

Since % means 'out of 100', to change a percentage to a decimal you just need to divide by 100.

$$45\% = 45 \div 100 = 0.45$$

Decimals to percentages

To change a decimal to a percentage, multiply by 100.

$$0.63 = 63\%$$

Decimals to fractions

Remember place value?

tenths column

⇩

0.92

⇧

hundredths column

So 0.92 can be written as $\frac{92}{100} = \frac{23}{25}$

NAIL IT!

Not sure what to use as the denominator of the fraction?

Look at where the last digit is within the decimal — this will tell you whether the denominator is 10, 100 or 1000 and so on.

For example, in 0.456 the last digit (6) is in the thousandths column, so $0.456 = \frac{456}{1000}$

Don't forget to cancel down if you can!

WORKIT!

Write 1.6 as a fraction.

1.6

$= 1\frac{6}{10}$ Don't forget the whole number part of the decimal.

$= 1\frac{3}{5}$

Always cancel the fraction down if you can.

Fractions to decimals

To change a fraction to a decimal, divide the numerator by the denominator.

For example, $\frac{3}{5} = 3 \div 5$

The numerator goes inside the division calculation. Add empty decimal places in case you need them.

Sometimes you will find the answer is a recurring decimal:

For example, $\frac{4}{7}$

As soon as you see the number has started to repeat you know it is a recurring decimal.

To indicate that a series of digits are recurring, place a dot over the first and last recurring digits:

$\frac{4}{7} = 0.\dot{5}7142\dot{8} = 0.571428571428571428\ldots$

Fractions to percentages

To change a fraction to a percentage, first change it to a decimal. Once you've written the fraction as a decimal, multiply by 100.

For example, $\frac{3}{5} = 0.6 = 60\%$

SNAP IT!

Fractions to percentages

$\frac{1}{2} = 0.5$	$= 50\%$
$\frac{1}{4} = 0.25$	$= 25\%$
$\frac{3}{4} = 0.75$	$= 75\%$
$\frac{1}{3} = 0.\dot{3}$	$= 33.\dot{3}\%$
$\frac{2}{3} = 0.\dot{6}$	$= 66.\dot{6}\%$
$\frac{1}{10} = 0.1$	$= 10\%$
$\frac{1}{5} = 0.2$	$= 20\%$
$\frac{1}{8} = 0.125$	$= 12.5\%$

DO IT!

Cut out six pieces of card the size of playing cards. On one side write the six titles:

Percentages to fractions *Decimals to fractions*

Fractions to decimals *Percentages to decimals*

Decimals to percentages *Fractions to percentages*

On the back of each card, write in your own words how you would convert them.

Spread the cards out in front of you with the titles showing. Test yourself to check you know how to carry out all the conversions – you could even play with a friend and make it a competition!

NAIL IT!

To compare fractions, decimals and percentages, write them all in the same form first.

CHECK IT!

Complete these questions without using a calculator.

1 Convert to fractions

  a 0.32   **c** 33%

  b 1.24   **d** 95%

2 Convert to decimals

  a $\frac{5}{12}$ **c** 49%   **e** $\frac{3}{7}$

  b $\frac{3}{8}$ **d** 18.5%

3 Convert to percentages

  a 0.91   **c** $\frac{4}{5}$

  b 0.3   **d** $\frac{9}{15}$

4 What percentage of the shape is shaded?

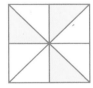

5 Place these quantities in order of size.

0.35 $\frac{2}{5}$ 30%

6 Amy correctly answered 15 out of 20 questions in a test. Rudi achieved 78%. Whose mark was higher? Show your workings.

Ordering fractions, decimals and percentages

Ordering fractions

To compare the size of fractions give them a **common denominator**.

To find the common denominator find the **lowest common multiple** of the denominators.

> See page 10 for more on lowest common multiples.

WORKIT!

Write in ascending order

$\frac{7}{10}$ $\frac{3}{5}$ $\frac{2}{3}$ $\frac{13}{20}$

$\frac{7 \times 6}{10} = \frac{42}{60}$ $\frac{3 \times 12}{5} = \frac{36}{60}$ $\frac{2}{3} = \frac{40}{60}$ $\frac{13}{20} = \frac{39}{60}$

$\frac{3}{5}$ $\frac{13}{20}$ $\frac{2}{3}$ $\frac{7}{10}$

> The lowest common multiple of 10, 5, 3 and 20 is 60.

> Compare the numerators of the rewritten fractions to give you the correct order for the original fractions.

Ordering fractions, decimals and percentages

Follow these steps when ordering fractions, decimals and percentages.

1. Split into < 0 and > 0.

2. Convert all the numbers into the same type of number, either fractions, decimals or percentages, to make them easy to compare.

NAILIT!

Use your common sense: if you know that −5.4 is smaller than $-\frac{1}{2}$ you don't need to convert $-\frac{1}{2}$ to a decimal.

< 0: −5.4, $-\frac{1}{2}$
> 0: $\frac{3}{4}$, 17%, 0.4
Writing all these numbers as decimals gives: −5.4, −0.5, 0.75, 0.17, 0.4

WORKIT!

Write in **ascending** order −5.4, $\frac{3}{4}$, 17%, $-\frac{1}{2}$, 0.4.

In ascending order:

$-5.4, \ -\frac{1}{2}, \ 17\%, \ 0.4, \ \frac{3}{4}$

> Don't forget to write the original numbers in order.

CHECKIT!

$\frac{8}{24}$ $\frac{9}{24}$ $\frac{14}{24}$

1. Write in **descending** order

$\frac{1 \times 8}{3}$ $\frac{3 \times 3}{8}$ $\frac{7 \times 2}{12}$ 14/24 9/24, 8/24

2. If these numbers were written in order, which would be in the middle?

−2.2   7 $\frac{1}{5}$ $-\frac{1}{10}$ 15%   1%   0.1   1%

−2.2, $-\frac{1}{10}$,

3. Jenny says, 'If the numerator of a fraction is less than half of the denominator I know the number is smaller than 0.5'.
Is she correct? Give a reason for your answer.

Calculating with fractions

Adding and subtracting fractions

In order to be able to add or subtract fractions you need them to have a common denominator.

If fractions already have a common denominator you can add or subtract them by adding or subtracting the numerators.

$$\frac{3}{5} + \frac{1}{5} = \frac{3+1}{5} = \frac{4}{5} \qquad\qquad \frac{9}{11} - \frac{7}{11} = \frac{9-7}{11} = \frac{2}{11}$$

If the fractions don't have a common denominator you need to get them in this form before adding or subtracting them.

$$\frac{2}{3} + \frac{1}{12} = \frac{8}{12} + \frac{1}{12}$$

> Find the lowest common multiple of the denominators and use this as the common denominator. Here it is 12.

$$= \frac{9}{12}$$

$$= \frac{3}{4}$$

> Always cancel your answer down.

A number containing a whole number (integer) and a fraction is called a **mixed number**.

To convert a mixed number to an improper fraction, multiply the whole number part by the denominator and add this to the existing numerator.

For example in $2\frac{3}{5}$ the whole number part is worth $\frac{(2 \times 5)}{5}$ so:

$$2\frac{3}{5} = \frac{10+3}{5} = \frac{13}{5}.$$

A fraction whose numerator is larger than the denominator is called an **improper fraction**.

To convert an improper fraction to a mixed number, divide the numerator by the denominator to give a whole number and a remainder. Write down the integer first, then write the remainder over the original denominator.

For example, to write $\frac{23}{20}$ as a mixed number, divide 23 by 20 to give 1 remainder 3: write 1 and put the 3 over 20.

$$\frac{23}{20} = 1\frac{3}{20}$$

WORK**IT!**

To add or subtract mixed numbers, first write them as improper fractions.

Work out $2\frac{2}{5} - 1\frac{1}{4}$

$$2\frac{2}{5} - 1\frac{1}{4} = \frac{12}{5} - \frac{5}{4}$$

> Write the numbers as improper fractions.

$$= \frac{48}{20} - \frac{25}{20}$$

> Find a common denominator.

$$= \frac{23}{20}$$

> Add or subtract the numerators.

$$= 1\frac{3}{20}$$

> Cancel down and rewrite as a mixed number.

STRETCHIT!

Do you always need to write the fractions as improper fractions before adding?

E.g. what is the best way to work out $1\frac{3}{5} + 2\frac{1}{4}$?

SNAPIT!   Adding or subtracting fractions

1 Write mixed numbers as improper fractions.

2 Give the fractions a common denominator.

3 Add or subtract the numerators.

4 Cancel down and write as a mixed number if necessary.

Multiplying fractions

To multiply fractions, work out: $\dfrac{\text{numerator} \times \text{numerator}}{\text{denominator} \times \text{denominator}}$

To multiply mixed numbers:

1 First write them as improper fractions.

$$1\frac{2}{3} \times \frac{4}{5} = \frac{5}{3} \times \frac{4}{5}$$

[handwritten: $1\,2/3$ $3+2=5/3 \times 4/5$]

2 Multiply the numerators and the denominators.

$$= \frac{5 \times 4}{3 \times 5} = \frac{20}{15}$$

[handwritten: $5 \times 4 = 20$ $3 \times 5 = 15$]

3 Cancel down and write as a mixed number.

$$= \frac{4}{3} = 1\frac{1}{3}$$

[handwritten: $1\,1/3$]

WORKIT!

Calculate $\dfrac{3}{4} \times \dfrac{6}{7}$

[handwritten: $3 \times 6 = \frac{18}{28} \Rightarrow \frac{2}{14}$]

$$\frac{3 \times 6}{4 \times 7} = \frac{18}{28}$$

$$= \frac{9}{14}$$ ← Make sure you cancel down.

Cross cancelling

When multiplying fractions you can simplify the calculation by cross cancelling first. This means dividing the numerator of one fraction and the denominator of the other by the same amount.

DOIT!

Multiply $\frac{3}{20} \times \frac{10}{21}$ without cross cancelling, then cancel your answer down. Now try again, but cross cancel first – you should get the same answer. Which method was easier?

WORKIT!

Work out $\dfrac{3}{20} \times \dfrac{10}{21}$

[handwritten: $3 \times 10 = \frac{30}{420}\ \frac{30}{14}$ 20×21]

$$\frac{3}{\underset{2}{20}} \times \frac{\overset{1}{10}}{21}$$ ← Divide one numerator and one denominator by 10.

$$\frac{\overset{1}{3}}{2} \times \frac{1}{\underset{7}{21}}$$ ← Divide one numerator and one denominator by 3.

$$\frac{1}{2} \times \frac{1}{7} = \frac{1}{14}$$

Dividing fractions

Dividing is the inverse of multiplying, so to divide fractions flip the second fraction and multiply.

To divide mixed numbers:

1 First write them as improper fractions.

$$2\frac{2}{3} \div \frac{3}{4} = \frac{8}{3} \div \frac{3}{4}$$

2 Flip the second fraction upside down and multiply.

$$= \frac{8}{3} \times \frac{4}{3} = \frac{32}{9}$$

3 Cancel down if necessary and write as a mixed number.

$$= 3\frac{5}{9}$$

WORKIT!

Calculate $\frac{3}{4} \div \frac{1}{2}$

$= \frac{3}{4} \times \frac{2}{1}$ — Flip the 2nd fraction and multiply.

$= \frac{6}{4}$

$= 1\frac{2}{4}$

$= 1\frac{1}{2}$ — Write your answer as a mixed number.

DOIT!

Write down two different mixed numbers.

For example, $2\frac{3}{5}$ and $1\frac{2}{3}$

Practise adding, subtracting, multiplying and dividing them. You can check your answers using the fraction key on your calculator.

Finding fractions of quantities

SNAPIT!

Finding a fraction of a number

To find a fraction of a number, divide by the denominator, then multiply by the numerator.

WORKIT!

Find $\frac{2}{3}$ of £12.45.

Divide by the denominator, 3.

£12.45 ÷ 3 = £4.15

£4.15 × 2 = £8.30 — Then multiply by the numerator, 2.

NAILIT!

Multiplying by $\frac{1}{a}$ is the same as dividing by a.

For example multiplying by $\frac{1}{5}$ is the same as dividing by 5.

CHECKIT!

1 Work out

a $2\frac{3}{8} - \frac{3}{4}$

b $\frac{15}{17} \times \frac{2}{5}$

c $\frac{1}{7} \times 3\frac{1}{3}$

d $2\frac{2}{5} + 5\frac{3}{4}$

e $\frac{1}{5} \div 2\frac{1}{2}$

2 Find

a $\frac{2}{5}$ of 30

b $\frac{7}{8}$ of £40

c $\frac{4}{9}$ of 1818 mm — Here 'of' means the same as ×.

3 In a class of 35 students, $\frac{3}{7}$ are studying French. How many students are not studying French?

4 Adrian thinks of a number. It is smaller than 50, and $\frac{2}{5}$ of it is a whole number greater than 12. What is the smallest number it could be?

Percentages

DOIT!

Which simpler percentages would you work out to find each of these?

30%   45%   80%

You may find there is more than one method!

Calculating a percentage of a value

You need to be able to find a percentage with and without a calculator. Remember that per cent (%) means 'out of 100'.

To find a percentage without a calculator, work out simpler percentages and add them together.

For example, to find 75% you could work out 50% and 25% and add them together.

SNAPIT!

Percentages without a calculator

You can work out some percentages without a calculator:

50% – halve it

25% – halve it then halve it again

10% – divide by 10

5% – find half of 10%

1% – divide by 100

WORKIT!

Find 37% of £16.00.

£16.00 ÷ 10 = £1.60   ← First find 10%.

£1.60 ÷ 2 = £0.80   ← Then find 5%.

£16.00 ÷ 100 = £0.16   ← Next find 1%.

37% = (3 × 10%) + 5% + (2 × 1%)

     = (3 × £1.60) + £0.80 + (2 × £0.16)

     = £4.80 + £0.80 + £0.32

     = £5.92

SNAPIT!

Percentages with a calculator

To find a percentage with a calculator, write it as a decimal, then multiply.

For example, to find 37%, multiply by 0.37.

NAILIT!

One of the most common mistakes with percentage increase or decrease is forgetting to add or subtract from the original price.

Since the price is being reduced, subtract from the original value.

Increasing or decreasing by a percentage

There are two ways to increase or decrease by a given percentage.

Method 1

Work out the increase and add it to the original value, or work out the decrease and subtract it from the original value.

WORKIT!

In a sale, a tablet is reduced by 20%. The original price of the tablet was £399. Work out the sale price.

£399 × 0.20 = £79.80   ← First calculate 20%.

£399 – £79.80 = £319.20

Method 2

Work out the multiplier. For example, if a value is increased by 20% then you need to find 120% of the original value.

120% = 1.2 so multiply the value by 1.2

If the value is decreased by 20%, then you need to find 80% of the original value.

80% = 0.8 so multiply the value by 0.8

WORKIT!

A shop reduces its prices by 10%. If the original price of a laptop computer is £380, work out the sale price.

0.9 × 380 = £342 ← If the price is reduced by 10% you need to find 90% of the original price, so multiply by 0.9.

DOIT!

Draw a flow diagram to show how to increase or decrease by a given percentage.

CHECKIT!

1 Without a calculator, work out

 a 10% of 18 cm 1.8

 b 25% of £1.20

 c 2% of 200 ml

2 Increase

 a 30 by 10%

 b 500 by 8%

 c £91 by 12%

3 Decrease

 a 600 by 20%   c £18 by 19%

 b 140 by 5%

4 In 2016 a college had 2800 students. By 2017 the number of students had increased by 9%. How many students attended the college in 2017?

5 A car loses 35% of its value in the first year. What is the value of a 1-year-old car that originally cost £22 000?

Handwritten annotations:

4 without calculator
① work out 9%
② Add it to original total
① 2800 × 0.09
 = 252
 (2800 − 28)
② 2800 + 252
 = 3052

or work out multiplier with calculator
 = you need to increase
by 9% so find 109 of
original value so
 multiple by
 1·09
 2800
 × 1·09
 = 3052

⑤ Find 35% and take
 away OR find
 multiplier g 100 − 35 = 65%
35% = 110 − 22,000

£22000

Order of operations

When performing calculations, you must follow the **order of operations** — the priority in which you work things out.

When the numerator and the denominator of a fraction are separate calculations, work them out separately before dividing.

Addition.

Brackets.

Finally, carry out the division.

WORKIT!

Calculate $\dfrac{3 \times 4^2 + 12}{10 - (1 + 1)}$

$3 \times 4^2 + 12$ ← Indices first.

$= 3 \times 16 + 12$

$= 48 + 12$ ← Multiplication.

$= 60$

$10 - (1 + 1)$

$= 10 - 2$

$= 8$ ← Subtraction.

$\dfrac{60}{8} = 7\dfrac{1}{2}$

SNAPIT!

BIDMAS

Use **BIDMAS** to remember the correct order of operations.

Brackets
Indices
Division
Multiplication
Addition
Subtraction

NAILIT!

Although we call it BIDMAS, multiplication and division can happen in any order. Addition and subtraction must be carried out by working from left to right, so sometimes you might do the subtraction first.

Using a calculator

Your calculator will follow the order of operations as long as you type in the correct calculation.

Find these calculator keys

Use these to indicate that you wish to carry out this part of the calculation first.

( ) brackets

x^2 square a number

$x^\square$ raise a number to a power

$\sqrt{}$ find the square root of a number

$(-)$ enter a negative number

$\sqrt[\square]{x}$ find a root of a number

ANS use the previous answer in a calculation

S-D change a number from an exact value to a decimal

$\dfrac{a}{b}$ or $\dfrac{\square}{\square}$ enter a fraction

DO IT!

Without a calculator, work out

$2 + 5.5 \times 3 - 16 \div 2^2 - (3.1 + 0.9)$

Now check it on your calculator – make sure you type it in exactly as it is written!

✓ CHECKIT!

1 Without a calculator, work out

  a $3 - 4 + 8$

  b $0.9 + 3.2 - \sqrt{35 + 1}$

  c $(5 - 6)^2 - 7 \times 2$

2 Use a calculator to work out

$$\frac{3.2^3 - \sqrt{1.4 - 0.99}}{2 - 8.6 \div 9.2}$$

Give your answer to 2 significant figures.

3 Put a pair of brackets into this calculation to make it as large as possible.

$8 - 3 + 5 \times 4$

Exact solutions

You must be able to work out exact solutions without using a calculator. To give an exact answer you may need to include π or fractions as part of the answer.

When adding multiples of π, treat it in the same way as simplifying an algebraic expression.

For example: $3π + 7π = 10π$

WORKIT!

Work out the exact area of

a a semicircle with diameter 12 cm

$$\text{Area of semicircle} = \tfrac{1}{2} \times \text{area of circle}$$

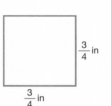 Area of a circle = π × r²

$$= \tfrac{1}{2} \times π \times r^2$$
$$= \tfrac{1}{2} \times π \times 6^2$$
$$= 18π \text{ cm}^2$$

12 cm

b a square with sides of length $\frac{3}{4}$ inch (in).

$$\text{Area} = \tfrac{3}{4} \times \tfrac{3}{4} = \tfrac{9}{16} \text{ in}^2$$

$\frac{3}{4}$ in

$\frac{3}{4}$ in

NAILIT!

Always sketch a shape if you are not given a picture, labelling any lengths you know. This will help you visualise the problem.

✓ CHECK**IT!**

1 Write as a single multiple of π

   a $2π - π$ **c** $\frac{1}{2}π + 2π$

   b $12π + 24π$

2 Find the exact solution to

   a $(3 + 4) \times π$ **b** $\frac{1}{8}π + \frac{1}{2}π$

3 Work out the exact area and perimeter of this shape.

$\frac{3}{4}$ cm

$\frac{2}{7}$ cm

4 Work out the exact value of each of these.

   a The circumference of a circle with radius 9 cm.

   b The area of a circle with radius 12 cm.

5 A square has perimeter equal to the circumference of a circle with radius 1 cm. Work out the exact length of one of the sides of the square.

Indices and roots

Indices

Indices are sometimes called 'powers'. An index tells you how many times to multiply a number by itself. For example, 5^4 means $5 \times 5 \times 5 \times 5$.

WORKIT!

1 Write the following in index form

  a $5 \times 5 \times 5 \times 5 \times 5 \times 5$

     5^6

  b 9×9

     9^2

2 Work out the value of

  a 3^2

     $3 \times 3 = 9$

  b 3^4

     $3 \times 3 \times 3 \times 3 = 81$

  c $(-1)^3$

     $-1 \times -1 \times -1 = -1$

Reciprocals

The reciprocal of a number is 1 divided by that number.
The reciprocal of 4 is $\frac{1}{4}$. It can be written as 4^{-1}.

A number multiplied by its reciprocal is 1. For example, $3^{-1} \times 3 = 1$.

The inverse of multiplying by a number is to multiply by the reciprocal.
For example, the inverse of multiplying by 4 is multiplying by $\frac{1}{4}$.

Negative indices

A negative index means 1 divided by that number (the reciprocal of the number). For example, 3^{-2} means the reciprocal of $3^2 = \frac{1}{3^2} = \frac{1}{9}$.

WORKIT!

Find the reciprocal of

0.125 becomes the denominator of a unit fraction.

a 12

  $\frac{1}{12}$ ← **1 divided by the number (making a unit fraction).**

     $\frac{8}{1}$ can be simplified to 8.

b 0.125

  $\frac{1}{0.125}$

  $= \frac{8}{1}$

  $= 8$

A fraction should not have a decimal as either numerator or denominator so multiply both top and bottom by 8.

SNAPIT!

Laws of indices

$$a^x \times a^y = a^{(x+y)}$$

$$a^x \div a^y = a^{(x-y)}$$

$$(a^x)^y = a^{xy}$$

Special indices

Any number to the power of 1 is itself. $7^1 = 7$

1 to any power is 1. $1^{12} = 1$

Any number to the power of 0 is 1. $3^0 = 1$

— 16th August 2017

Laws of indices

These rules can be used to simplify expressions.

SNAPIT!

Special indices

$$a^{-1} = \frac{1}{a} \qquad a^1 = a$$

$$a^{-x} = \frac{1}{a^x} \qquad a^0 = 1$$

$$a^{-1} \times a = 1$$

WORKIT!

a Simplify the expression $4^2 \times 4^{-5}$, leaving your answer in index form.

$$4^2 \times 4^{-5} = 4^{(2+\,-5)} = 4^{-3}$$

b Work out the value of $\frac{(3^7)^2}{3^{11}}$

$$\frac{(3^7)^2}{3^{11}} = \frac{3^{7\times 2}}{3^{11}} = \frac{3^{14}}{3^{11}} = 3^{14-11} = 3^3 = 3 \times 3 \times 3 = 27$$

NAILIT!

When you need to find the square root of a number there are two possible solutions, negative and positive.

For example,
$-7 \times -7 = 49$ **and**
$7 \times 7 = 49$

In some problems, only the positive solution is meaningful and relevant, so think about what the numbers refer to.

It is either 6.01 or 6.3. 6.01^2 is very close to 36 therefore a better estimate is 6.3.

Square roots

The square root is the inverse of squaring a number. The square root sign is $\sqrt{\ }$

$\sqrt{49}$ means 'the number that multiplies by itself to give 49'.

$\sqrt{49} = 7$ or -7 ◄ Two negatives make a positive when multiplied.

If the exact root of a number cannot be found then it can be left written as a root. This is called a **surd** — for example, $\sqrt{7}$.

The approximate root of a number can be found by considering which two square numbers it lies between.

WORKIT!

Which of the values shown is closest to $\sqrt{40}$?

6.01   7.2   5.9   6.3

We know that: $\sqrt{36} = 6$ and $\sqrt{49} = 7$

Therefore: $6 < \sqrt{40} < 7$.

6.3

The square root of a larger number can be found by writing it as a product of prime factors. For example, to find the square root of 225:

Write as a product of prime factors:

$225 = 3^2 \times 5^2$.

Remember that the square root is the inverse of squaring, so: $\sqrt{225} = \sqrt{3^2 \times 5^2} = 3 \times 5 = 15$.

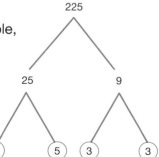

Other roots

The sign for other roots is $\sqrt[a]{\ }$

$\sqrt[3]{8}$ means 'the number that multiplies by itself 3 times to make 8'.

$\sqrt[4]{10\,000}$ means 'the number that multiplies by itself 4 times to make 10 000'.

a is a positive integer, e.g. 3 or 4

NAILIT!

Any root of 1 is 1.

For example, $\sqrt[15]{1} = 1$.

WORKIT!

1 Work out $\sqrt{64}$

$8 \times 8 = 64$ and

$-8 \times -8 = 64$

So $\sqrt{64} = 8$ or -8

2 Calculate

a $\sqrt[3]{8}$

$2 \times 2 \times 2 = 8$

So $\sqrt[3]{8} = 2$

b $\sqrt[10]{1}$

$1 \times 1 \times 1 \times 1 \times 1 \times 1$
$\times 1 \times 1 \times 1 \times 1 = 1$

So $\sqrt[10]{1} = 1$ or -1

If you replaced every 1 with −1 the answer would also be 1. This is because an **even** number of minus signs, multiplied together, give you a positive answer.

To estimate the cube root of any positive number, consider the nearest integers (higher and lower) that have known cube roots. For example, to find the range within which $\sqrt[3]{50}$ lies, think about the cube numbers:

Cube numbers: 1   8   27   64   125 …

50 lies between 27 and 64, therefore:

$$\sqrt[3]{27} < \sqrt[3]{50} < \sqrt[3]{64}$$
$$3 < \sqrt[3]{50} < 4$$

STRETCHIT!

Now you know that the cube root of 50 is between 3 and 4, you could test 3.5 x 3.5 x 3.5 to see whether this is an overestimate or an underestimate. Then you could use your answer to test a new value and keep going until you get an answer very close to 50.

CHECKIT!

1 Find the reciprocal of

  a 3   **b** 0.4   **c** 0.9

2 Write these four numbers in order of size, from smallest to largest.

  3^2 1^3 $\sqrt[3]{27}$ $\sqrt[3]{8}$

3 Work out

  a $(-2)^3$ **b** 1^{12} **c** $(-9)^2$ **d** 17^0

4 What is the value of

  a 4^{-1} **c** 1^{-4}

  b 7^{-2} **d** the reciprocal of 3?

5 Simplify the expression $\dfrac{5^9}{5^2 \times 5^3}$ leaving your answer in index form.

6 Write in ascending order

  $\sqrt{20}$ 3.7 $\dfrac{1}{12}$ $\sqrt[3]{7}$

7 Work out the square root of 196 by writing 196 as the product of prime numbers.

— Start 21st August 2017.

Standard form

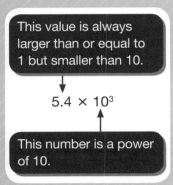

This value is always larger than or equal to 1 but smaller than 10.

5.4×10^3

This number is a power of 10.

Standard form is used to write very large and very small numbers. Numbers in standard form are written in two parts, a numerical value (1 or more, but less than 10) and a power of 10. For example:

5.4×10^3.

The number can be written in 'ordinary form' by multiplying out.

$5.4 \times 10^3 = 5.4 \times 10 \times 10 \times 10$

$= 5400$

If the power of 10 is negative you can read it as dividing by the power of 10.

$3.2 \times 10^{-3} = 3.2 \div (10 \times 10 \times 10)$

$= 0.0032$

WORKIT!

a Work out $(2.5 \times 10^3) + (4.2 \times 10^5)$.

$(2.5 \times 10 \times 10 \times 10) + (4.2 \times 10 \times 10 \times 10 \times 10 \times 10)$

$= 2500 + 420\,000$

$= 422\,500$

b Write 9.5×10^{-5} as an ordinary number.

$9.5 \div (10 \times 10 \times 10 \times 10 \times 10)$

$= 0.000095$

SNAPIT!   Standard form

To write 0.00000567 in standard form:

1. Put the decimal point after the first **non-zero** digit.

   5.67

2. Work out how many times you have multiplied or divided by 10 and use this as the power of 10.

   6 times, so use 10^6 *10 × 10 × 10 × 10 × 10 × 10*

3. If the original number is very large then the index is positive. If it is very small then the index is negative.

   5.67×10^{-6}

Calculations involving standard form

To multiply or divide numbers written in standard form use the laws of indices.

For example: $(3.2 \times 10^5) \times (5 \times 10^9)$ can be rewritten as 1.6×10^{15}.

$3.2 \times 5 \times 10^5 \times 10^9 = 16 \times 10^{14}$ ◄──── Combine the powers of 10.

$16 \times 10^{14} = 1.6 \times 10^{15}$

Divide 16 by 10 (to give 1.6) and multiply 10^{14} by 10 (to give 10^{15}), so that your answer is written correctly in standard form.

NAIL IT!

Remember, multiplication is commutative. This means the order in which you carry it out does not matter: $2 \times 3 = 3 \times 2$.

WORKIT!

Work out the value of $(2.5 \times 10^9) \div (5 \times 10^4)$. Give your answer in standard form.

Rewrite the calculation as a fraction: $\dfrac{2.5 \times 10^9}{5 \times 10^4}$

Cancel the powers of 10 using the laws of indices: $= \dfrac{2.5 \times 10^5}{5}$

Divide by 5: $= 0.5 \times 10^5$

Rewrite in standard form: $= 5 \times 10^4$

✓ CHECKIT!

1 Write these numbers as ordinary numbers.

  a 4.5×10^7 **b** 9.1×10^{-2}

2 Write these numbers in standard form.

  a 645 000 000   **b** 0.000000079

3 Work out the value of
  $(3.5 \times 10^5) - (4.2 \times 10^3)$

4 Write these four numbers in order of size from smallest to largest.

  3.2×10^2 3.1×10^{-2}
  3.09×10 $3 + (2.1 \times 10^2)$

5 The speed of light is approximately 300 000 000 m/s. Write this in standard form.

6 A piece of A4 paper is 1.1×10^{-4} metres thick. Work out the height of a stack of 200 sheets.

Listing strategies

STRETCH IT!

You could alternatively start by considering each colour at a time. How would the list look then?

red +
small red +
medium red +
small green

To work out the total number of combinations of different events you must use strategies to avoid missing out or repeating outcomes.

For example, if a shop sells T-shirts in three different sizes (small, medium and large) and three different colours (red, green and blue), you could consider the different options for one size at a time, as shown below.

small + red	small + green	small + blue
medium + red	medium + green	medium + blue
large + red	large + green	large + blue

Sample space diagrams

You can draw up a sample space diagram to help you list the possible outcomes of an event.

For the shop selling T-shirts, draw up a table with the size options horizontally and the colour options vertically.

	small	**medium**	**large**
red	red + small	red + medium	red + large
green	green + small	green + medium	green + large
blue	blue + small	blue + medium	blue + large

Drawing up a sample space can help prevent you missing out any combinations.

See page 179 for more on sample space diagrams.

DO IT!

A dice with numbers 1 to 6 and a coin with sides showing heads and tails are thrown.

List all the different possible outcomes.

Then draw up a sample space to show all the possible outcomes. Which method works better?

1 – 6 and heads & tails

1 – 6
2

dice has more options so would work better

CHECK IT!

1 List all the different three-digit numbers that can be made using the digits 1, 2 and 3.

2 List all the different numbers smaller than 500 that can be me using the digits 4, 6 and 9.

3 A factory manufactures shampoo in three different sizes: small, medium and large. They make four different types of shampoo: almond, briar rose, chamomile and detox. List all the different products the factory produces.

You may **not** use a calculator for these questions.

1 Write down a prime number and a factor of 12 that add up to 13.

2 Write 620 as a product of prime factors using indices.

3 Find the highest common factor of 18, 36 and 40.

4 Write these numbers in ascending order:
$$-3.2 \quad -11.5 \quad 1.4 \quad -8.3 \quad -3.5$$

5 Calculate
 a £32.99 + £18.74 **c** Use inverse operations
 b £54.99 ÷ 3   to check these calculations.

6 Work out **a** 23 × 0.14 **b** 27 × 149

7 Work out $\frac{1}{3}$ of 81.

8 Work out 345 ÷ 11, giving your answer as a mixed number.

9 **a** Write $\frac{3}{8}$ as a decimal.
 b Write 0.7 as a percentage.

10 Give your answer to each of these calculations in its simplest form.
 a Write 70% as a fraction.
 b Write 0.8 as a fraction.

11 Which of these fractions is larger than $\frac{1}{4}$?
 $\frac{1}{2}$ $\frac{2}{7}$ $\frac{3}{11}$ $\frac{2}{5}$

12 Give your answer to each of these calculations in its simplest form.
 a $\frac{3}{5} + \frac{1}{7}$ **b** $2\frac{1}{5} - \frac{7}{10}$ **c** $\frac{2}{3} \div \frac{4}{9}$

13 Which of these is larger?
 $0.25 - 0.07$ or $\frac{2}{3} - \frac{1}{2}$

14 Work out $\frac{3}{5} \times 1\frac{1}{4}$.

15 Write 0.045 as a fraction. Give your answer in its simplest form.

16 Find 25% of £8.60. $= £2.15$

17 Work out **a** $(-3)^2$ **b** $\sqrt[3]{125}$

18 Between which two integers does $\sqrt{30}$ lie?

19 Write each of the following in standard form
 a 3 400 000 000 **b** 0.000000304

20 The length of a journey is measured as 37.6 miles to the nearest tenth of a mile.
 Write down the error interval for the length of the journey.

21 Here are three cards.
 a Write down the largest two-digit multiple of 3 that can be made using the cards.
 Each card may be used only once.
 b Write down all the two-digit numbers that can be made using these cards.

 | 2 | 5 | 1 |

22 A unit of electricity costs 9.086p. On average a family uses 242.6 units a month.
 a Work out an estimate for the cost of electricity for a year.
 b Is your answer an overestimate or an underestimate? Give a reason for your answer.

23 At a school of 600 students, 40% of the students cycle to school and $\frac{1}{5}$ walk. The rest travel by bus. How many travel by bus?

24 Rhiannon's result in a test is less than 50 out of 100. Her mark is a multiple of 5. She answered more than $\frac{1}{3}$ of the questions correctly. What is the smallest value her mark could be?

25 Barbara says, 'The sum of three different prime numbers will always be odd.'
 Is she correct? Give a reason for your answer.

26 Work out the value of the following, giving your answer in standard form.
 a $(4 \times 10^8) \times (2.7 \times 10^2)$
 b $(1.8 \times 10^5) \div (6 \times 10^2)$

You **may** use a calculator for these questions.

27 In a sale, a musical instrument that normally costs £349 is reduced by 20%. What is the new price of the instrument?

28 Round 3.05092 to
 a 1 decimal place **b** 3 significant figures.

29 Round 324 567 to **a** the nearest thousand
   b 2 significant figures.

30 Find the value of 3^6.

31 Four people share the cost of an outing equally between them. The costs are:
 Group travel card £26.25
 Picnic   £18.23
 4 tickets at   £5.50 each
 How much does each person pay?

32 19% of people in a town own a cat. The population is 18000. How many people own a cat?

33 The table gives information about the price (in pence) of fuel in the UK.

2010	2011	2012	2013	2014	2015	2016
111.4	127.8	133.4	132.9	130.1	108.2	102.3

 a Between which two years was the greatest fuel price increase?
 b Between 2016 and 2017 the price increased by 10%. What is the price of petrol in 2017?

Algebra

Understanding expressions, equations, formulae and identities

DOIT!

Draw a poster to highlight the differences between expressions, equations, formulae and identities. Use your own words and include an example of each.

In algebra you use letters to represent unknown numbers.

$4a + 2b$ is an **expression**.

It does not have an equals (=) sign. The parts that are separated by + or − are called **terms**. In $4a + 2b$ the terms are $4a$ and $2b$. An expression can contain letter terms and/or number terms.

$2x + 2 = 8$ is an **equation**.

It has an equals sign. It contains letter terms and numbers. You can solve an equation to find the value of the letter. An equation is only true for certain values of the letter. Here, the value of x is 3.

$A = lw$ is a **formula** (plural: **formulae**).

It has an equals sign. The letters represent different quantities. The letters are **variables**, as their values can vary. You can use a formula to calculate one variable if you know the other variables. For example, you can use $A = lw$ to find A if you know l and w.

$\frac{10x}{2} \equiv 5x$ is an **identity** (plural: **identities**).

An identity is true for all values of the letters. Here, the two sides of $\frac{10x}{2} \equiv 5x$ are equal for all values of x. An identity can be written with the identity symbol '$\equiv$' or with an equals sign.

WORKIT!

Write down whether each of these is an expression, an equation, a formula or an identity.

a $A = \frac{1}{2}bh$

Formula ← The value of A can be worked out if the values of b and h are known.

b $5x + 1 = 16$

Equation ← It can be solved to find the value of x.

c $3x + 2y + 4x = 7x + 2y$

Identity ← Collecting the like terms on the left-hand side gives $7x + 2y$ which is equal to the right-hand side. (See page 44 for more on collecting like terms.)

d $2m - 4$

Expression ← It does not have an equals sign.

CHECKIT!

1 $c = \pi d$ $3a + 6 = 10$
$3ab + 2ab = 5ab$ $3(a + 2)$

From the above list, write down:

  a an equation   **c** an expression

  b a formula   **d** an identity.

2 Victor says that $4x - 2 = 2x$ is an identity. James says that $4x - 2 = 2x$ is an equation.

Who is correct? Give a reason for your answer.

Simplifying expressions

For an unknown number x, the **expression** $2x + 3$ means '2 lots of x with 3 added' or 'add 3 to 2 lots of x'. '2 lots of x' is the same as '$2 \times x$' or '$x + x$' and is written as $2x$.

Expressions that involve $\times$ and $\div$ should be simplified where possible.

Multiplying expressions

To multiply together two or more algebraic expressions, first multiply any number parts and then multiply the letters.

 SNAPIT! **Algebraic notation**

$3a$ is used for $3 \times a$
ab is used for $a \times b$
a^2 is used for $a \times a$
a^3 is used for $a \times a \times a$
a^2b is used for $a \times a \times b$
$\dfrac{a}{b}$ is used for $a \div b$

WORKIT!

Simplify

a $a \times a$

a^2

> Use indices to show a letter multiplied by itself. a^2 is read as 'a squared'.

b $b \times b \times b$

b^3

> Use indices to show the same letter multiplied together three times. b^3 is read as 'b cubed'.

c $2 \times j \times k \times 6$

$12jk$

> Multiply the numbers together: $2 \times 6 = 12$
> Multiply the letters together by writing them next to each other: $j \times k = jk$
> Write the number in front of the letters.

d $2d \times 4d$

$8d^2$

Dividing expressions

Dividing algebraic expressions is similar to cancelling fractions. If necessary, write the division as a fraction first. Cancel any number parts where possible and then cancel any letters where possible.

> Write the division as a fraction.
> Cancel the number parts by dividing the top and bottom by 2.

WORKIT!

Simplify

a $6a \div 2$

$\dfrac{\overset{3}{\cancel{6}}a}{\cancel{2}} = 3a$

b $12xy \div 4y$

$\dfrac{\overset{3}{\cancel{12}}x\cancel{y}}{\cancel{4}\cancel{y}} = 3x$

> This is the same as $\dfrac{12 \times x \times y}{4 \times y}$.
> Cancel the number parts by dividing the top and bottom by 4.
> y appears on the top and bottom, so cancel.

 CHECKIT!

1 Simplify

a $p \times p \times p$

b $4 \times b \times c \times 7$

c $4a \times 3b$

d $5x \times 4x$

e $2g \times (-4g)$

f $2p \times 3q \times r$

2 Simplify

a $10x \div 2$

b $\dfrac{14w}{-2}$

c $6p \div p$

d $8mn \div 2m$

e $\dfrac{12xy}{3y}$

f $9abc \div bc$

Collecting like terms

NAILIT!

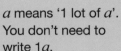

a means '1 lot of a'. You don't need to write $1a$.

You can simplify algebraic expressions by collecting **like terms**.

Like terms contain the same letter, or combination of letters, with the same powers.

a, $4a$ and $-10a$ are all like terms.

b^2, $2b^2$ and $8b^2$ are all like terms.

$2a$, $2ab$ and $2a^2$ are **not** like terms.

ab^2 and a^2b are **not** like terms.

You can add and subtract like terms. To do this you need to consider the $+$ or $-$ sign in front of each term.

WORKIT!

Simplify

a $p + p + p + p$

   $4p$ ◀ — 4p means '4 lots of p' or '$4 \times p$' or '$p + p + p + p$'.

b $3x + x - 2x$

   $2x$ ◀ — x means $1x$. So there are $3 + 1 - 2$ lots of x.

When an expression contains two or more different letters, first identify the like terms. This also applies when expressions contain combinations of letters and powers.

WORKIT!

Simplify

a $4a + 2b + 2a - 4b$ — $4a$ and $2a$ are like terms. $2b$ and $4b$ are like terms.

   $6a - 2b$ ◀

b $2f + 3g - f + 4g + 5$ —   First collect the f terms together and the g terms together. Then deal with the numbers on their own.

   $f + 7g + 5$ ◀

You can also simplify expressions that involve surds. If you're unsure about surds have a look at page 36.

$2\sqrt{x} + 5\sqrt{x} = 7\sqrt{x}$ $\qquad\qquad$ $8\sqrt{x} - 3\sqrt{x} = 5\sqrt{x}$

You can see that $2\sqrt{x}$ means '2 lots of $\sqrt{x}$' or '$2 \times \sqrt{x}$' or '$\sqrt{x} + \sqrt{x}$'.

DOIT!

Choose two different letters – for example, a and b. On a set of plain cards, write down different terms using those letters, including $+$ and $-$ signs in front of them. Example terms: $3a$, b^2, $2ab$, $-2b$.

Shuffle the cards and deal out 2, 3, 4 or 5 cards so that they form an expression. Simplify each expression where possible.

CHECKIT!

1 Simplify

a $f + f + f + f + f$

b $8b - 3b + 2b$

c $4mn + 2mn - mn$

d $4a + 6 - a - 5$

e $3d + 4e + d - 6e$

f $2x + 5y + 3x - 2y - 2$

g $3a - 2b + 4a + 7b$

h $2a - b - 5a - 3$

i $x^2 + x^2$

j $2t^3 + 4 - t^3 - 4$

k $a + a + b \times b$ ◀ — Look carefully at the signs!

l $4\sqrt{x} + 3\sqrt{x}$

m $7\sqrt{x} - 4\sqrt{x}$

n $12\sqrt{x} - \sqrt{x} - 4\sqrt{x}$

Using indices

The terms a^3 and b^4 are written using index notation.

The 3 in a^3 and the 4 in b^4 is called a **power** or **index** (plural: **indices**). It tells you how many times the letter must be multiplied by itself.

a^3 means $a \times a \times a$ and b^4 means $b \times b \times b \times b$.

The a in a^3 and the b in b^4 is called the **base**.

You can simplify algebraic expressions involving powers by using the laws of indices.

You can only use the laws of indices when the bases are the same.

To multiply powers of the same base, add the indices.

WORKIT!

Simplify

a $x^4 \times x^3$

$x^4 \times x^3 = x^{4+3} = x^7$ ◀ — Add the indices. $x^m \times x^n = x^{m+n}$

b $2f^2 \times 3f^5$

$2f^2 \times 3f^5 = 2 \times 3 \times f^2 \times f^5 = 6 \times f^{2+5} = 6f^7$ ◀ — Group the numbers and the letters together. Multiply the numbers 2 and 3. Add the indices.

c $4x^2y^3 \times 3xy^2$

$4x^2y^3 \times 3xy^2 = 4 \times 3 \times x^2 \times x \times y^3 \times y^2$

$= 12 \times x^{2+1} \times y^{3+2} = 12x^3y^5$ ◀ — Group the numbers (4 and 3) and the same letters (x^2 and x, y^3 and y^2) together.

NAILIT!

It's always a good idea to show your working.

To divide powers of the same base, subtract the indices.

WORKIT!

Simplify

a $x^9 \div x$ ◀ — If there is no index, the base has a power of 1, so $x = x^1$.

$x^9 \div x = x^{9-1} = x^8$ ◀ — Subtract the indices. $x^m \div x^n = x^{m-n}$

b $\dfrac{10p^6}{5p^4}$

$\dfrac{10p^6}{5p^4} = (10 \div 5) \times (p^6 \div p^4) = 2 \times p^{6-4} = 2p^2$ ◀ — Group the numbers and the letters together. Divide the numbers, then subtract the indices.

c $4x^5 \div 16x^3$

$4x^5 \div 16x^3 = \dfrac{4x^5}{16x^3} = \dfrac{4}{16} \times \dfrac{x^5}{x^3} = \dfrac{4}{16} \times x^{5-3} = \dfrac{x^2}{4}$ ◀ — Write the division in fraction form. Simplify the numerical fraction: $\frac{4}{16} = \frac{1}{4}$. Leave the final answer in fraction form (rather than using the decimal form $0.25x^2$).

45

SNAPIT!

Laws of indices

$x^m \times x^n = x^{m+n}$

$x^m \div x^n = x^{m-n}$

$(x^m)^n = x^{m \times n}$

$x = x^1$

$x^0 = 1$

$x^{-m} = \dfrac{1}{x^m}$

To raise a power of a base to a further power, multiply the indices.

WORKIT!

Simplify

a $(x^4)^3$

$(x^4)^3 = x^{4 \times 3} = x^{12}$

> Multiply the indices.
> $(x^m)^n = x^{m \times n}$

b $(2y^3)^2$

$(2y^3)^2 = 2^2 \times (y^3)^2 = 4 \times y^{3 \times 2} = 4y^6$

> Both 2 and y^3 are raised to the power of 2, or squared.
>
> Work out 2^2, then multiply the indices.

Anything raised to the power 0 is equal to 1. So $m^0 = 1$ and $y^0 = 1$.

A negative power means 'one over'.

WORKIT!

Simplify

a a^{-1}

$\dfrac{1}{a}$

> A term raised to a power -1 is 'one over' the term.
> $x^{-1} = \frac{1}{x}$

b b^{-3}

$\dfrac{1}{b^3}$

> Use $x^{-m} = \dfrac{1}{x^m}$

c $c^{-3} \times c^{-2}$

$c^{-3} \times c^{-2} = c^{-3+(-2)} = c^{-5} = \dfrac{1}{c^5}$

> Add the indices.
> $x^m \times x^n = x^{m+n}$
> Be careful with the negative signs:
> $+ -2 = -2$
> Give the final answer in simplified form without the negative power.

d $(d^{-3})^{-2}$

$(d^{-3})^{-2} = d^{-3 \times (-2)} = d^6$

> Multiply the indices.
> $(x^m)^n = x^{m \times n}$
> negative × negative = positive

CHECKIT!

1 Simplify

a $x^5 \times x^4$

b $p \times p^4$

c $2m^4 \times 3m^4$

d $3m^4n \times 5m^2n^3$

e $u^{-2} \times u^5$

f $t^7 \times t^{-6}$

2 Simplify

a $x^4 \div x^2$

b $\dfrac{y^7}{y^3}$

c $\dfrac{p^9}{p^8}$

d $8x^6 \div 4x^3$

e $m^3 \div m^5$

f $\dfrac{5x^8}{15x^4}$

g $3x^2 \div 9x$

3 Simplify

a $(x^2)^3$

b $(y^4)^4$

c $(p^5)^2$

d $(4m^5)^2$

e $(x^2)^{-3}$

f $(n^{-4})^{-2}$

4 Simplify fully $\dfrac{x^3 \times x^5}{x^4}$

> First add the indices on the top.

Expanding brackets

Expand single brackets

To **expand** a bracket, multiply **every** term inside the brackets by the term outside the brackets.

$3 \times x = 3x$

$3(x + 2) = 3x + 6$

$3 \times 2 = 6$

Here, there are **two** multiplications to be done when expanding a single bracket and you will obtain **two** terms. If you're unsure about multiplying expressions, go to page 43.

WORKIT!

Expand

a $4(2a + 3)$

$4(2a + 3) = 8a + 12$ ◄——— $4(2a + 3) = (4 \times 2a) + (4 \times 3)$

b $b(b - 4)$

$b(b - 4) = b^2 - 4b$ ◄——— $b(b - 4) = (b \times b) - (b \times 4)$

c $2c(c + 5)$

$2c(c + 5) = 2c^2 + 10c$ ◄——— $2c(c + 5) = (2c \times c) + (2c \times 5)$

To 'expand and simplify', expand the brackets and then collect like terms. Collecting like terms is covered on page 44.

$2 \times a = 2a \quad 4 \times 2a = 8a$

$2(a + 4) + 4(2a - 3) \quad = 2a + 8 + 8a - 12$
$= 2a + 8a + 8 - 12$
$2 \times 4 = 8 \quad 4 \times -3 = -12 \quad = 10a - 4$

Each term includes the + or − sign in front of it. So in the first bracket, the number term is +4 and in the second bracket it is −3.

You need to take care when there is a − sign outside the brackets.

WORKIT!

Expand and simplify

$2(3y + 1) - 2(2y - 3)$

$2(3y + 1) - 2(2y - 3)$

$= 6y + 2 - 4y + 6$ ◄———

$= 2y + 8$

Make sure you answer the question asked. Here you are told to 'expand and simplify' so you need to multiply out the brackets ('expand') and then collect like terms ('simplify').

DOIT!

In your own words, make a note of what is meant by these instructions.

1 Expand

2 Expand and simplify

3 Collect like terms

STRETCHIT!

Expand

a $a(\sqrt{3} + a)$

b $b(\sqrt{5} - b)$

c $(\sqrt{c})^2 + (\sqrt{d})^2$

Remember: '$2\sqrt{x}$' means '2 lots of $\sqrt{x}$' or '$2 \times \sqrt{x}$'

SNAPIT!

Negative terms outside brackets

When one term is − and one term is + the product is −

$-3 \times p = -3p$

$-3(p + 4) = -3p - 12$

$-3 \times 4 = -12$

When one term is − and the other term is − the product is +

$-2 \times q = -2q$

$-2(q - 3) = -2q + 6$

$-2 \times -3 = 6$

Expand double brackets

To **expand** two brackets, multiply **every** term inside one bracket by **each** term in the other bracket.

Here there are **four** multiplications to be done when expanding two brackets and you will obtain **four** terms before you collect like terms to simplify the expression.

$x \times 3 = 3x$

$x \times x = x^2$

$(x + 2)(x + 3) = x^2 + 3x + 2x + 6$

$2 \times x = 2x$ $\qquad = x^2 + 5x + 6$

$2 \times 3 = 6$

NAILIT!

$(a + b)^2$ is **not** equal to $a^2 + b^2$.

SNAPIT! Expanding brackets

To expand two brackets
$(x + 2)(x + 3)$ use **FOIL**

First term $(x + 2)(x + 3)$ x^2

Outer term $(x + 2)(x + 3)$ $+ 3x$

Inner term $(x + 2)(x + 3)$ $+ 2x$

Last term $(x + 2)(x + 3)$ $+ 6$

You can use the acronym **FOIL** to help you expand two brackets.

You need to be able to: 'Expand and simplify $(x + 2)^2$'. Here you are being asked to square a single bracket. First you need to write out $(x + 2)^2$ in full: $(x + 2)(x + 2)$. Then expand and simplify as normal.

WORKIT!

Expand and simplify $(x - 3)^2$

$(x - 3)^2 = (x - 3)(x - 3)$ ◀ First write out $(x - 3)^2$ in full.

$= x^2 - 3x - 3x + 9$

$= x^2 - 6x + 9$

Use FOIL to find the four terms of the expansion. Then collect like terms as the question is 'expand and simplify'.

Take care with the negative signs:
$x \times -3 = -3x$ $\qquad -3 \times -3 = 9$

STRETCHIT!

1 Find the missing terms in this expression.
$(x + 2)(x + \square) = x^2 + 6x + \square$

2 Expand and simplify
  a $(x + 3)(2x + 2)$ **c** $(2x + 3)(3x - 1)$
  b $(3x - 2)(x + 4)$ **d** $(2x - y)(3x + y)$

CHECKIT!

1 Expand
  a $3(a + 2)$ **e** $4(x + y + 2)$
  b $4(b - 4)$ **f** $-(2y + 4)$
  c $5(2c + 5)$ **g** $x(x - 2)$
  d $2(3 - e)$ **h** $2a(a + 5)$

2 Simplify
  a $6a - (3a + 5)$
  b $4x - 6 + 2(x + 5)$

3 Expand and simplify
  a $2(2x + 3) + 4(x + 5)$
  b $3(3y + 1) + 2(4y - 3)$
  c $4(2m + 4) - 3(2m - 5)$

4 Expand and simplify
  a $(x + 2)(x + 3)$ **c** $(a + 3)(a - 7)$
  b $(y - 3)(y + 4)$ **d** $(m - 1)(m - 6)$

5 Expand and simplify
  a $(x + 1)^2$ **c** $(m - 2)^2$
  b $(x - 1)^2$ **d** $(y + 3)^2$

Factorising

Taking out common factors with single brackets

Factorising is the opposite of expanding brackets.
You need to put brackets in.

To factorise an expression	$3a + 12$
① Find the highest common factor of the terms in the expression.	3 is the highest common factor of the terms $3a$ and 12.
② Write the factor outside the brackets.	$3a + 12 = 3(\qquad)$
③ Work out what you need to multiply the factor by to get each term in the original expression.	$3 \times a = 3a$ and $3 \times 4 = 12$ $3a + 12 = 3(a \qquad 4)$
④ Write the + or − sign from the original expression inside the brackets.	$3a + 12 = 3(a + 4)$

Note that 3 is a **factor** of $3a + 12$. Also, $a + 4$ is a factor of $3a + 12$.

Always check your answer by expanding the brackets (see page 47) to see if you get back to the original expression.

$3 \times a = 3a$

$3(a + 4) = 3a + 12$

$3 \times 4 = 12$

SNAP IT! Factorising

expanding brackets

$2(a + 3) \xrightarrow{\qquad} 2a + 6$

factorising

NAIL IT!

The common factor can be a number or a letter or both.

Always check that your answer cannot be factorised further.

WORKIT!

Factorise

a $2a + 6$

$2a + 6 = 2(a + 3)$

Take 2 outside the bracket. It is the highest common factor of 2 and 6.

b $8b - 12$

$8b - 12 = 4(2b - 3)$

The common factors of 8 and 12 are 2 and 4. The highest common factor is 4.

$2(4b - 6)$ is **not** fully factorised.

c $5xy + 15x$

$5xy + 15x = 5x(y + 3)$

The common factors of $5xy$ and $15x$ are x and $5x$. The highest common factor is $5x$.

$x(5y + 15)$ is **not** fully factorised.

d $c^2 + 8c$

$c^2 + 8c = c(c + 8)$

c is the common factor of c^2 ($c \times c$) and $8c$ ($8 \times c$).

e $ad + bd$

$ad + bd = d(a + b)$

d is the common factor of ad and bd.

DOIT!

Identify all the quadratic expressions in this sub-topic up to this point.

Factorising quadratic expressions

A **quadratic expression** has a squared term (x^2) and no higher power.

The following are examples of quadratic expressions of the form $x^2 + bx + c$ where b and c are numbers.

$$x^2 + 4x + 4 \qquad x^2 + 3x - 18 \qquad x^2 - 6x + 5 \qquad x^2 - 2x - 15$$

To factorise a quadratic expression of this type you need to write the expression with two brackets: $(x \quad)(x \quad)$. An x in each bracket will give x^2 when multiplied.

To factorise the quadratic expression $x^2 + 5x + 6$ you need to find two numbers that multiply to make $+6$ and add up to $+5$:

1 Identify number pairs that multiply to make 6:

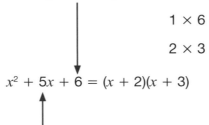

$$1 \times 6$$
$$2 \times 3$$

$$x^2 + 5x + 6 = (x + 2)(x + 3)$$

> For two numbers to multiply to make $+6$ and add up (sum) to $+5$, both numbers must be positive.

2 Which number pair adds up to 5?

$$1 + 6 = 7 \ \text{✗}$$
$$2 + 3 = 5 \ \text{✓}$$

3 Write each number with the corresponding sign ($+2$, $+3$) in each of the brackets with x.

4 Check your answer by expanding the brackets to see if you get back to the original expression.

$$(x + 2)(x + 3) = x^2 + 3x + 2x + 6$$
$$= x^2 + 5x + 6$$

> If you're unsure about expanding double brackets, have a look at page 48.

WORKIT!

Factorise $x^2 + 2x - 3$

$x^2 + 2x - 3$

Number pairs that multiply to make -3:

$1 \times -3, \ -1 \times 3$

Sum of number pairs:

$1 + -3 = -2 \ \text{✗}$

$-1 + 3 = 2 \ \text{✓}$

$x^2 + 2x - 3 = (x - 1)(x + 3)$

> For two numbers to multiply to make -3, one number must be negative and one must be positive.

WORKIT!

Factorise $x^2 - 6x + 5$

$x^2 - 6x + 5$

Number pairs that multiply to make 5:

$1 \times 5, -1 \times -5$

Sum of number pairs:

$1 + 5 = 6$ ✗

$-1 + -5 = -6$ ✓

$x^2 - 6x + 5 = (x - 1)(x - 5)$

> For two numbers to multiply to make $+5$ and add up to -6, both numbers must be negative.

SNAPIT!   Difference of two squares

Factorise quadratic expressions of the form $x^2 - a^2$ using the rule:

$x^2 - a^2 \equiv (x + a)(x - a)$

STRETCHIT!

The area of a rectangle is $x^2 + 3x + 2$.

The length of the rectangle is $x + 2$.

Find an expression for the width of the rectangle.

WORKIT!

Factorise $x^2 - 25$

$x^2 - 25 = x^2 - 5^2$

> 25 is a square number: $25 = 5^2$
> Use $x^2 - a^2 = (x + a)(x - a)$ with $a = 5$

$= (x + 5)(x - 5)$

✓ CHECKIT!

1 Factorise

a $3a + 9$

b $5b - 10$

c $7 + 14c$

d $d^2 - 2d$

2 Factorise

a $8a + 20$

b $4b - 12$

c $18 + 9c$

d $2d^2 - 3d$

3 Factorise

a $4x - 6y$

b $am + bm$

c $12a^2 + 8a$

d $4x^2 + 3xy$

e $2n - 9n^2$

f $5x + 10xy$

g $4pq - 12p$

h $4x^2y - 8y$

4 The expression $4(x - 3) + 3(2x + 6)$ simplifies to $a(5x + b)$.

Work out the values of a and b.

5 Factorise

a $x^2 + 8x + 7$

b $x^2 + 4x - 5$

c $x^2 - 2x - 8$

d $x^2 - 5x + 6$

e $x^2 - 6x + 9$

f $x^2 + 7x + 12$

g $x^2 + 3x - 10$

h $x^2 - x - 20$

6 Factorise

a $x^2 - 16$

b $x^2 - 36$

c $x^2 - 81$

d $y^2 - 100$

Substituting into expressions

SNAPIT! BIDMAS

Use **BIDMAS** to remember the correct order of operations.

Brackets
Indices
Division
Multiplication
Addition
Subtraction

$x + y$, $2a + 3$, $4mn$, $3x^2 - 2$, $3(p + 1)$ and $5pq + pqr$ are all examples of algebraic expressions. The letters are used to represent unknown numbers.

If you are given the value of each of the letters, you can **substitute** them into the expression to work out the value of the expression.

You need to use the correct order of operations when doing the calculation. If you're unsure about this, see page 32.

The value of the letters may be positive or negative.

Use brackets when substituting a negative number.
$5p$ means $5 \times p$.
If $p = 2$ then
$5p = 5 \times 2 = 10$
Work out each multiplication first (BIDMAS).

WORKIT!

1 Work out the value of $5p - 2q$ when $p = 2$ and $q = -3$

$5 \times 2 - 2 \times (-3) = 10 - (-6)$

$= 10 + 6$ ← Subtracting a negative number is the same as adding.

$= 16$

2 Work out the value of $r(s - 4)$ when $r = 5$ and $s = -4$

$5 \times (-4 - 4) = 5 \times -8$

$= -40$ ← Work out the brackets first (**BIDMAS**).
'positive' × 'negative' = 'negative'

Take care when substituting negative numbers into expressions involving powers.

NAILIT!

Substitute all the values **before** doing any calculations.

WORKIT!

1 Work out the value of $m^2 + n^2$ when $m = 3$ and $n = -1$

$3^2 + (-1)^2 = 9 + 1$ ← Work out each power (indices) first (BIDMAS).
$(-1)^2 = -1 \times -1 = 1$

$= 10$

'negative' × 'negative' = 'positive'

2 Work out the value of $5x^3$ when $x = -2$

$5 \times (-2)^3 = 5 \times -8$

$= -40$

Only the value of x is cubed: $5x^3$ is not the same as $(5x)^3$.
$(-2)^3 = -2 \times -2 \times -2$
$= -8$
'negative' × 'negative' × 'negative' = 'negative'

You will also need to be able to substitute decimals and fractions.

WORKIT!

1 Work out the value of $12ab$ when $a = \frac{1}{4}$ and $b = 2$

$12 \times \frac{1}{4} \times 2 = 6$

> Work from left to right. $12 \times \frac{1}{4}$ is the same as $12 \div 4 \ (= 3)$

2 Work out the value of $4c + 8d$ when $c = 0.5$ and $d = 2.25$

$(4 \times 0.5) + (8 \times 2.25) = 2 + 18 = 20$

> Use brackets around each multiplication.
> $0.5 = \frac{1}{2}$ and $2.25 = 2\frac{1}{4}$
> $8 \times 2\frac{1}{4} = (8 \times 2) + \left(8 \times \frac{1}{4}\right) = 16 + 2$
> If you can use a calculator for this question, simply input the numbers in decimal form.

Care should also be taken when the expression is in fraction form.

WORKIT!

Work out the value of $\frac{4(x - 2y)}{3}$ when $x = 5$ and $y = 7$

$\frac{4(5 - 2 \times 7)}{3} = \frac{4(5 - 14)}{3}$

> To work out the brackets you first need to do the multiplication.

$= \frac{4 \times -9}{3}$

$= \frac{-36}{3} = -12$

> Here, the multiplication 4×-9 has been done before the division. An alternative method would be to first divide the top and bottom by 3:
> $\frac{4 \times -9}{3} = 4 \times -3 = -12$

CHECKIT!

1 Work out the value of $5a + 2b$ when $a = 3$ and $b = -2$.

2 Work out the value of each expression when $x = 2$ and $y = -4$.

  a $x - 2y$ **d** $x^2 + y^2$

  b $3xy$ **e** $2x + 4(x - y)$

  c $4y - 3x$ **f** $\frac{1}{2}(x + y)$

3 True or false: $3a^2 = 81$ when $a = 3$?

Give a reason for your answer.

4 Work out the value of each expression when $p = \frac{1}{2}$ and $q = -4$.

  a $10pq$ **c** $\frac{q}{p}$

  b $8p^2$ **d** $2q^2 - 12p$

5 Work out the value of $\frac{d(e - 2)}{f}$ when $d = 7$, $e = -3$ and $f = 10$.

Writing expressions

For an unknown number x, the expression $x + 3$ means '3 more than x' or 'x with 3 added' or 'add 3 to x'.

To write expressions correctly you need to use the notation and symbols with care.

> If you're unsure about algebraic notation, have a look at page 43.

When writing expressions to solve problems or represent real-life situations, the expression should be given in its simplest form.

DOIT!

1 Draw a square with sides of length x. Write down an expression for its perimeter and area. Repeat for a rectangle with length 2 and width $x + 2$.

2 Draw a triangle with angles a, $a - 20$, $a + 50$. Write down an expression for the sum of its angles.

WORKIT!

1 Use algebra to write down an expression for each of these numbers.

  a 4 less than a

    $a - 4$

> $a - 4$ means '4 less than a'. $a - 4$ is **not** the same as $4 - a$.

  b a multiplied by 2 and then 4 added

    $2a + 4$

  c a plus 4 and then multiplied by 2

    $2(a + 4)$

> Use a bracket to show that '$a + 4$' is multiplied by 2.
> $2a + 4$ and $2(a + 4)$ are **not** the same.

WORKIT!

The width of a rectangle is x cm. Its length is 3 times its width.

Write down an expression for its perimeter.

  Length $= 3x$

> First write down an expression for the length. $3x$ means '3 times x'.

  Perimeter $= x + 3x + x + 3x$

          $= 8x$ cm

> Simplify the expression by collecting like terms (see page 44).

> Perimeter is the distance around the shape. You are adding 2 widths and 2 lengths so you could also write the expression as:
> $2 \times x + 2 \times 3x$
> $= 2x + 6x = 8x$

✓ CHECKIT!

1 Use algebra to write down an expression for each of these numbers.

  a q less than 4   **d** p multiplied
  b m more than n by itself
  c x multiplied by y

2 Nathan buys x white T-shirts and y patterned T-shirts for his holiday. Write down an expression for the total number of T-shirts he buys.

3 £y is shared equally between eight people. How much does each person receive?

4 Idris sells vintage jewellery. The necklaces cost £100 and the bracelets cost £75. One day he sold n necklaces and b bracelets. Write an expression for the total amount of money, in pounds, that he received for these sales.

5 A triangle has sides of length $3a$, $2a + 4$ and $4a - 2$ cm. Write down an expression for its perimeter.

6 Here is a right-angled triangle. Write down an expression for its area.

$2a + 5$

4

Solving linear equations

Solving simple equations

$2x + 2 = 8$ is a **linear equation**. An equation is only true for certain values of the letter. You can **solve** the equation to find the value of the letter.

To solve an equation you need to rearrange it so that the letter is on its own on one side of the equation. To rearrange an equation you can:

- add the same number to both sides

  $a - 4 = 12$

  $a = 12 + 4$ ← Add 4 to both sides.

  $a = 16$

- subtract the same number from both sides

  $b + 8 = 13$

  $b = 13 - 8$ ← Subtract 8 from both sides.

  $b = 5$

- divide both sides by the same number

  $4c = 20$ ← $4c$ means $4 \times c$

  $c = \dfrac{20}{4}$ ← Divide both sides by 4.

  $c = 5$

- multiply both sides by the same number

  $\dfrac{d}{8} = 3$ ← $\dfrac{d}{8}$ means $d \div 8$

  $d = 3 \times 8$ ← Multiply both sides by 8.

  $d = 24$

> **NAILIT!**
>
> The solution to an equation can be a positive number, negative number, fraction or decimal.

When solving an equation with two operations, look to do the $+$ or $-$ first.

WORKIT!

Solve

a $4x - 2 = 8$

$4x = 10$ ← Add 2 to both sides.

$x = 2.5$ ← Divide both sides by 4. The answer can be given as a decimal ($x = 2.5$) or as a mixed number $\left(x = 2\tfrac{1}{2}\right)$.

b $\dfrac{p}{6} + 2 = 7$

$\dfrac{p}{6} = 5$ ← Subtract 2 from both sides.

$p = 30$ ← Multiply both sides by 6.

c $\dfrac{x + 5}{2} = 6$

$x + 5 = 12$ ← Multiply both sides by 2.

$x = 7$ ← Subtract 5 from both sides.

> You need to do two operations to get x on its own. Do them one at a time, dealing with the '$- 2$' first.

Take care when the **coefficient** (the number in front of the letter) is negative.

> The first step is to make the coefficient of x positive.

> Subtract 5 from both sides.

WORKIT!

Solve $13 - 4x = 5$

$13 - 4x = 5$

$13 = 5 + 4x$ ← Add $4x$ to both sides.

$8 = 4x$

$2 = x$ ← Divide both sides by 4. You can write the answer as '$2 = x$' or '$x = 2$'.

> **DOIT!**
>
> Solve the Work it! equation $13 - 4x = 5$ using these alternative steps:
>
> Step 1: subtract 13 from both sides.
>
> Step 2: divide both sides by -4.

Solving equations with brackets and the unknown on both sides

NAIL IT!

When solving equations:
- show all your working
- use a new line for each step of working
- do the same thing to **both** sides of the equation.

To solve an equation with brackets, expand the brackets first. If you're unsure about expanding brackets, have a look at page 47.

$2(x + 3) = 7$

$2x + 6 = 7$ ← $2(x + 3) = 2 \times x + 2 \times 3$

$2x = 1$ ← Subtract 6 from both sides.

$x = \dfrac{1}{2}$ ← Divide both sides by 2.

WORKIT!

Solve $2(3m + 1) + 6 = 2$

$2(3m + 1) + 6 = 2$

$6m + 2 + 6 = 2$ ← Expand brackets.

$6m + 8 = 2$ ← Collect like terms.

$6m = -6$ ← Subtract 8 from both sides.

$m = -1$ ← Divide both sides by 6.

Both x terms ($3x$ and x) have a positive coefficient. Subtract the smaller term in x from both sides so that the coefficient of x remains positive.

In an equation where the letter appears on both sides, the first step is to rearrange the equation so that the letter appears on **one** side only.

$3x + 9 = x + 5$

$2x + 9 = 5$

$2x = -4$ ← Subtract 9 from both sides.

$x = -2$ ← Divide both sides by 2.

NAIL IT!

When solving linear equations with the unknown on both sides, collect the terms in the unknown on the side of the equation that gives them a positive coefficient.

WORKIT!

Solve $5 - 3x = 7 - 4x$

$5 - 3x = 7 - 4x$ ← Both x terms ($-3x$ and $-4x$) have a negative coefficient. Add the higher term in x ($4x$) to both sides so that the coefficient of x becomes positive.

$5 + x = 7$

$x = 2$

Subtract 5 from both sides.

WORKIT!

Solve $4(2x + 1) = 3(5x - 1)$

$4(2x + 1) = 3(5x - 1)$

$8x + 4 = 15x - 3$ ← Expand the brackets.

$4 = 7x - 3$ ← Subtract $8x$ from both sides.

$7 = 7x$ ← Add 3 to both sides.

$1 = x$ ← Divide both sides by 7.

DOIT!

Create a mind map for 'Solving linear equations'. Include an example of all the different equation types and illustrate a worked example for each type with all the key points.

CHECKIT!

1 Solve

  a $5a = 35$

  b $b - 9 = 8$

  c $\frac{c}{4} = 4$

  d $d + 4 = 2$

2 Solve

  a $2x + 3 = 13$

  b $3y - 4 = 11$

  c $2p + 9 = 1$

  d $\frac{f}{3} - 7 = 4$

  e $\frac{x + 5}{2} = 8$

  f $\frac{f - 7}{3} = 4$

3 Solve

  a $9 - m = 7$

  b $10 - 3x = 1$

  c $7 - 2x = 2$

  d $5 = 1 - 2f$

4 Hannah solves the equation $2x + 4 = 8$

Here is her working:

$2x + 4 = 8$

$2x = 8$

$x = 4$

Hannah's answer is wrong.
Write down the mistake that she
has made.

5 Solve

  a $3(a + 2) = 15$

  b $4(b - 2) = 4$

  c $3(4c - 9) = 9$

  d $2(d + 3) + 4 = 2$

  e $4(2x + 3) - 2 = 6$

6 Solve

  a $3m = m + 6$

  b $5t - 6 = 2t + 3$

  c $4x + 3 = 2x + 8$

  d $3 - 2p = 6 - 3p$

  e $3y - 8 = 5y + 4$

7 Solve

  a $2(x + 5) = x + 6$

  b $7b - 2 = 2(b + 4)$

  c $4(2y + 1) = 3(5y - 1)$

  d $2x - 1 = 8 - 4x$

Writing linear equations

DO IT!

Make a note of these key facts which you may need when setting up equations to solve problems.

1. Sum of angles on a straight line and around a point.
2. Sum of angles in a triangle and in a quadrilateral.
3. Property of opposite angles of a parallelogram.
4. Properties of isosceles and equilateral triangles.
5. Length properties of a square and rectangle.

Using algebra to solve angle or perimeter problems

For some problems you will use the information given in the question to set up an equation.

WORKIT!

The diagram shows a quadrilateral.

All measurements are in centimetres.

a Write down an expression, in terms of x, for the perimeter of the quadrilateral.

$\text{Perimeter} = 3x + 2x + 3 + x + 2x - 7$

$= 8x - 4$

> Simplify the expression by collecting like terms (page 44). x means '1x'.

b The perimeter of the quadrilateral is 48 cm.

Find the value of x.

> You are told that the perimeter is 48 cm. Use this fact to turn your expression from part a into an equation. Solve your equation to find the value of x.

$8x - 4 = 48$

$8x = 52$

> Add 4 to both sides.

$x = 6.5 \text{ cm}$

> Divide both sides by 8.

For some problems you will need to use your knowledge of angles and properties of shapes to set up an equation.

WORKIT!

Here is a triangle.

Work out the size of the smallest angle.

> Write an equation in terms of a using the fact that the angle sum of a triangle is 180°.

$a + 20 + a + 25 + a - 15 = 180$

$3a + 30 = 180$

> Subtract 30 from both sides.

$3a = 150$

> Divide both sides by 3.

$a = 50$

$a - 15 = 50 - 15 = 35$

> Make sure you answer the question asked. The smallest angle is given by the expression $a - 15$.

The smallest angle is 35°.

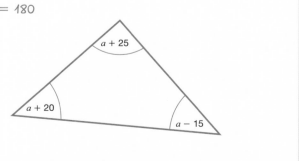

Using algebra to solve word problems

For some word problems you need to choose your own letter to represent the unknown. State the letter you have chosen and what it represents at the start of your working. For example:

'Imelda has 3 more sweets than Jack. Katerina has twice as many sweets as Imelda. They have 25 sweets in total. Work out how many sweets Imelda has.'

> You can use any letter to represent the unknown.

Let n = number of sweets that Jack has.

$n + n + 3 + 2(n + 3) = 25$

> Form an equation for the total number of sweets (which equals 25).
> Jack has n sweets.
> Imelda has $n + 3$ ('3 more' than Jack)
> Katerina has $2(n + 3)$ ('twice as many' as Imelda)

> Expand the brackets.

$n + n + 3 + 2n + 6 = 25$

> Collect like terms.

$4n + 9 = 25$

> Subtract 9 from both sides.

$4n = 16$

> Divide both sides by 4.

$n = 4$

> Substitute $n = 4$ to find the number of sweets Imelda has.

Jack has 4 sweets.

Imelda has $n + 3 = 4 + 3 = 7$ sweets.

✓ CHECK**IT!**

1 A square has sides of length $2s + 3$ cm.

  a Write down an expression, in terms of s, for the perimeter of the square.

  b The perimeter of the square is 84 cm. Find the value of s.

2 Here is a quadrilateral.

  a Work out the value of x.

  b Work out the size of the largest angle.

3 Monica is 4 years younger than Karen.

  The sum of their ages is 64.

  Find Monica's age.

4 Multiplying a number by 2 and then adding 4 gives the same answer as subtracting the number from 16.

  Find the number.

5 The width of a rectangle is 2 cm smaller than its length. Its perimeter is 36 cm.

  Find the width of the rectangle.

6 Here is an isosceles triangle.

  Work out the value of a and the value of b.

Linear inequalities

NAILIT!

The wide end (or 'mouth') of the inequality always points towards the larger value.

Picture a number line: −5 is to the left of −3.

Inequalities and number lines

Inequalities can be used to compare quantities. They are written using inequality signs.

4 > 2 is a true statement as 4 is greater than 2.

−5 < −3 is a true statement as −5 is less than −3.

For $a < b$ to be a true statement, the value of a would have to be less than the value of b.

SNAPIT!   **Inequality signs**

> means 'is greater than'

≥ means 'is greater than or equal to'

< means 'is less than'

≤ means 'is less than or equal to'

For $x \geq 2$ to be a true statement, the value of x would have to be greater than or equal to 2. So if x is an integer it could be 2 or 3 or 4 or 5 and so on.

An **integer** is a positive or negative whole number, including 0.

DOIT!

Choose a positive or negative integer for a. Choose a positive or negative integer for b. Draw number lines to show these inequalities. Label each inequality on your diagram.

1 $x > a, x < b$
2 $x \geq a, x \leq b$
3 $a < x < b$
4 $a < x \leq b$
5 $a \leq x < b$
6 $a \leq x \leq b$

WORKIT!

Write down all the integers that satisfy the inequality $-3 \leq x < 2$

$x = -3, -2, -1, 0, 1$

$-3 \leq x < 2$ means x is greater than or equal to −3 but less than 2. So −3 is included but 2 isn't.

An inequality can be shown on a number line.

$x > -2$

The empty (or open) circle shows that −2 is **not** included.

$x \leq 2$

The filled (or closed) circle shows that 2 **is** included.

WORKIT!

a Write down the inequality represented on the number line.

$x \geq -4$ ◄——— The circle is filled so −4 **is** included.

b Show the inequality $-2 < x \leq 3$ on a number line.

$-2 < x \leq 3$ means x is greater than −2 but less than or equal to 3. All the values between −2 and 3 must be shown on the number line.

−2 is not included so use an empty circle.

3 is included so use a filled circle.

Solving inequalities

You can solve inequalities in the same way that you solve linear equations.

Use the inequality sign on each line of your working.

You may need to represent the solution set on a number line.

If you're unsure about solving equations, have a look at page 55.

WORKIT!

a Solve the inequality $2x + 6 \geq 1$

Subtract 6 from both sides. Keep the inequality sign the same.

$2x + 6 \geq 1$

$2x \geq -5$

$x \geq -2\frac{1}{2}$

Divide both sides by 2. The answer could be written as $x \geq -\frac{5}{2}$.

b Represent the solution set on a number line.

NAILIT!

The solution to an inequality is written using an inequality sign (and **not** an equals sign).

You also need to be able to solve inequalities where the unknown appears on both sides of the inequality sign.

WORKIT!

Solve the inequality $3x - 4 < 5x$

$3x - 4 < 5x$　　Subtract $3x$ from both sides.

$-4 < 2x$　　Divide both sides by 2.

$-2 < x$

You also need to be able to solve two inequalities in x.

WORKIT!

Write down all the integers that satisfy the inequality $-9 \leq 3n < 12$

$-9 \leq 3n < 12$　　Divide each term in the inequality by 3.

$-3 \leq n < 4$

The integer solutions are $-3, -2, -1, 0, 1, 2, 3$.

Care must be taken when there is a negative sign in front of the unknown.

To solve $8 - 3x > 2$ you can choose between two methods.

The first method involves adding $3x$ to both sides to ensure the coefficient of x is not negative.

$8 - 3x > 2$

$8 > 2 + 3x$　　Add $3x$ to both sides.

$6 > 3x$　　Subtract 2 from both sides.

$2 > x$ or $x < 2$　　Divide both sides by 3.

The second method involves a division by −3.

$$8 - 3x > 2$$

Subtract 8 from both sides.

$$-3x > -6$$

Divide both sides by −3.

$$x < 2$$

You must reverse the inequality sign. (See Snap it! below.)

 SNAPIT!

When you multiply or divide both sides of an inequality by a negative number you must reverse the inequality sign.

CHECKIT!

1 Write down all the integers that satisfy these inequalities.

a $2 < x < 6$

b $2 \leq x < 6$

c $-1 < x \leq 3$

d $-3 \leq x \leq 1$

2 Write down the inequality represented on each number line.

a
```
 ←+——+——+——+——+——+——○——+→
  -3  -2  -1   0   1   2   3   4
```

b
```
 ←+——+——●——+——+——+——+——+→
  -4  -3  -2  -1   0   1   2   3
```

c
```
 ←+——+——●——+——+——+——+——●→
  -3  -2  -1   0   1   2   3   4   5
```

3 Show each inequality on a number line.

a $x \geq 1$

b $1 < x \leq 4$

c $-2 \leq x < 2$

d $-3 \leq x \leq 0$

4 Solve these inequalities. Represent each solution set on a number line.

a $2x - 2 > 4$

b $2x - 7 \leq 3$

c $4x + 3 \leq 13$

d $4x < 2x - 10$

e $4x - 8 < 6x$

f $7x + 2 \geq 3x - 2$

5 Write down all the integers that satisfy these inequalities.

a $-12 < 4x \leq 8$

b $-8 \leq 2x < 14$

c $-6 < 6x \leq 18$

d $9 \leq 3n \leq 15$

6 Solve these inequalities.

a $4 - x \leq 1$

b $6 - 3x > 9$

c $8 - 2x \geq 7$

d $-2 < -x \leq 3$

Formulae

Using formulae

A **formula** (plural: **formulae**) is a mathematical rule. It gives the relationship between different quantities.

width

length

The word formula for the area of this rectangle is:

area = length × width

The same formula can be written using algebra as: $A = lw$

You can use a formula to calculate one quantity if you know the other quantities. You can substitute the values of length (l) and width (w) into this formula to find the area (A).

DOIT!

Make a poster to show all the formulae that you need to know for your maths exam.

WORKIT!

Jazz uses this rule to work out her mobile phone bill:

total bill = monthly charge + cost per minute of calls × number of minutes of calls.

The monthly charge is £12. For calls in excess of her contract limit, the cost per minute is 7p.

Jazz made 80 minutes of calls above her contract limit.

Work out her total bill.

7p = £0.07 ← *The values substituted into a formula must be in the same units. Write 7p in pounds.*

Total bill = £12 + £0.07 × 80 ← *Substitute all the values before doing any calculations. Use the correct order of operations: **M**ultiplication before **A**ddition (BID**MA**S).*

　　　　= £12 + £5.60

　　　　= £17.60

WORKIT!

This formula can be used to work out the distance, s, moved by an object.

$s = ut + \frac{1}{2}at^2$　*The velocity of an object is its speed in a particular direction.*

where u is velocity, t is time and a is acceleration.

Work out s when $u = 5$, $t = 4$ and $a = 10$.

$s = 5 \times 4 + \frac{1}{2} \times 10 \times 4^2$ ← *Substitute the values of u, t and a into the formula.*

$= 5 \times 4 + \frac{1}{2} \times 10 \times 16$ ← *Work out the power (or indices) first (B**I**DMAS). Then work out the multiplications (BID**M**AS).*

$= 20 + 80$ ←

$= 100$　← *Do the addition last (BIDMA**S**).*

Use a new line for each step of working.

NAILIT!

To find a quantity using a formula:

* use the same units of measurement

* use the correct order of operations: BIDMAS (see page 32).

Writing formulae

You can use information given in words to write a word formula or an algebraic formula. This can then be used to solve a problem.

WORKIT!

This rule can be used to work out the cooking time of a turkey:

Cooking time in minutes = 15 × weight in kg + 70

A turkey weighs w kg and is cooked for T minutes.

Write down a formula for T in terms of w.

> 15w means '15 × w'.

$T = 15w + 70$

> You do not need to include the units in the algebraic formula.

WORKIT!

The cost of printing promotional leaflets is £150 plus 2p per leaflet.

The cost of printing n leaflets is C pounds.

Write down a formula for C in terms of n.

$2p = £0.02$

> The cost is in pounds so write 2p in pounds.

$C = 150 + 0.02n$

When writing a formula for the perimeter or area of a shape you should give the formula in its simplest form. To do this, you need to be able to collect like terms (page 44) and multiply expressions (page 43).

DOIT!

> A regular polygon has sides of equal length, l. Write down a formula for the perimeter, P, of a regular polygon with five sides.

> The perimeter is the distance around a shape.

WORKIT!

The diagram shows an isosceles triangle.

> $2x - 2$

> An isosceles triangle has two equal sides (as shown by the dashes on two of the sides).

> x

Write down a formula for the perimeter P of the triangle in terms of x.

> Write down all three sides before collecting like terms to simplify the formula.

$P = 2x - 2 + 2x - 2 + x$

$P = 5x - 4$

Rearranging formulae

The letter on its own on one side of the equals sign in a formula is called the **subject**. In the formula $A = lw$ the subject is A.

To find the value of a quantity that is not the subject of the formula, substitute the given values into the formula and then solve the equation.

If you're unsure about solving equations, see page 55.

WORKIT!

This formula can be used to work out the amount Tomasz the builder is paid for a job:

$P = 80 + 20h$

where P is the amount paid in pounds and h is the number of hours worked.

Tomasz is paid £520 for a job.

How many hours did he work?

$520 = 80 + 20h$

$440 = 20h$ ◄── Subtract 80 from both sides.

$22 = h$ ◄── Divide both sides by 20.

Tomasz worked 22 hours.

You can check your answer by substituting the value of h back into the formula:

$P = 80 + 20 \times 22$
$= 80 + 440$
$= 520 \checkmark$

To change the subject of a formula you have to **do the same thing to both sides** of the formula until you have the required letter on its own on one side.

WORKIT!

Rearrange each formula to make the letter in square brackets the subject.

a $y = mx + c$ $[m]$ ◄── You need to get m on its own on the right-hand side (RHS).

$y = mx + c$ ◄── First subtract the terms you don't need on the RHS ($-c$).

$y - c = mx$ ◄── mx means '$m \times x$' so divide by x to get m on its own.

$\dfrac{y - c}{x} = m$ ◄── Everything on the left-hand side (LHS) is divided by x.

b $A = \dfrac{1}{2}bh$ $[b]$

$A = \dfrac{1}{2}bh$

$2A = bh$ ◄── Multiply both sides by 2 to get rid of the fraction.

$\dfrac{2A}{h} = b$ ◄── Divide both sides by h.

c $a = \dfrac{b}{c} + d$ $[b]$

$a = \dfrac{b}{c} + d$ ◄── Deal with the '$+ d$' term first.

$a - d = \dfrac{b}{c}$

$c(a - d) = b$ ◄── Everything on the LHS is multiplied by c. You need to use brackets to show this.

d $A = \pi r^2$ $[r]$

$A = \pi r^2$

$\dfrac{A}{\pi} = r^2$ ◄── Divide both sides by π.

$\sqrt{\dfrac{A}{\pi}} = r$ ◄── Square root both sides to get r as the subject.

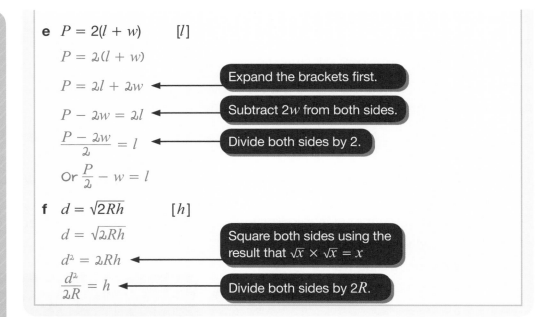

DO IT!

Make l the subject of the formula $P = 2(l + w)$ using this method.

1 Divide both sides by 2.

2 Subtract w from both sides.

Compare this method with that shown in Work it! part **e**.

e $P = 2(l + w)$ $[l]$

$P = 2(l + w)$

$P = 2l + 2w$ ← Expand the brackets first.

$P - 2w = 2l$ ← Subtract $2w$ from both sides.

$\dfrac{P - 2w}{2} = l$ ← Divide both sides by 2.

Or $\dfrac{P}{2} - w = l$

f $d = \sqrt{2Rh}$ $[h]$

$d = \sqrt{2Rh}$

$d^2 = 2Rh$ ← Square both sides using the result that $\sqrt{x} \times \sqrt{x} = x$

$\dfrac{d^2}{2R} = h$ ← Divide both sides by $2R$.

✓ CHECK IT!

1 Use the formula

wage earned =
hours worked × hourly rate + bonus
to calculate the wage William earns when he works for 35 hours at £8 an hour and receives a bonus of £25.

2 Convert 45° C to °F using the formula
$F = \dfrac{9}{5}C + 32$

3 This formula can be used to work out the final velocity, v, of an object:

$v = u + at$ where u is initial velocity, a is acceleration and t is time.

Work out the value of v when $u = 10$, $a = -20$ and $t = 5$.

4 This rule can be used to work out the cost, in pounds, of hiring a floor sander.

Multiply the number of days by 25 and then add 50.

Betty hires a floor sander for d days.

The cost is £C.

Write down a formula for C in terms of d.

5 This rule can be used to change a distance given in miles, m, to a distance in kilometres, k.

First multiply by 8 then divide by 5.

a Write this rule as a formula.

b Use your formula to change 200 miles into kilometres.

6 **a** Write down a formula for the perimeter P of the rectangle in terms of a.

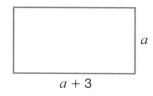
a
$a + 3$

b Work out the value of P when $a = 6$ cm.

7 $S = \dfrac{D}{T}$

Work out the value of D when $T = 6.5$ and $S = -10$.

8 Rearrange each formula to make the letter in square brackets the subject.

a $v = u + at$ $[a]$

b $V = \dfrac{1}{3} Ah$ $[h]$

c $y = 3(x - 3)$ $[x]$

d $v^2 = u^2 + 2as$ $[s]$

e $T = \sqrt{\dfrac{2s}{g}}$ $[g]$

Linear sequences

Number sequences

A pattern of numbers that follow a rule is called a **sequence**.

Each number in a sequence is called a **term**.

In the sequence 3, 6, 9, 12, ... the first term is 3, the second term is 6, the third term is 9, and so on.

You can continue a sequence of numbers if you know the **term-to-term rule**. This rule tells you what to do to each term to obtain the next term in the sequence.

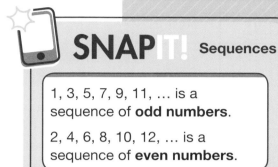

SNAPIT! **Sequences**

1, 3, 5, 7, 9, 11, ... is a sequence of **odd numbers**.

2, 4, 6, 8, 10, 12, ... is a sequence of **even numbers**.

WORKIT!

Here are the first four terms in a sequence:

46, 42, 38, 34

46   42   38   34
-4 -4 -4

a Write down the term-to-term rule of this sequence.

The term-to-term rule is subtract 4.

Work out the difference between **consecutive terms** (those next to each other in the sequence).

b Write down the next two terms in this sequence.

30, 26

c Write down the 10th term in this sequence.

7th term = 22

8th term = 18

9th term = 14

10th term = 10

Use your term-to-term rule from part a to find the next two terms:

$34 - 4 = 30$

$30 - 4 = 26$

Continue the number sequence until you get to the 10th term.

You can also continue a number sequence, or find a specific term, if you know the **position-to-term rule**. This rule tells you what to do to the term number to obtain that term in the sequence.

WORKIT!

The position-to-term rule for a sequence is 'multiply the term number by 2 and add 4'.

a Find the first two terms of this sequence.

1st term = $1 \times 2 + 4 = 6$

The 1st term is term number 1, the 2nd term is term number 2.

2nd term = $2 \times 2 + 4 = 8$

The first two terms are 6, 8.

b Find the 100th term of this sequence.

100th term = $100 \times 2 + 4 = 204$

Pattern sequences

Here are the first four patterns in a sequence of patterns made with squares.

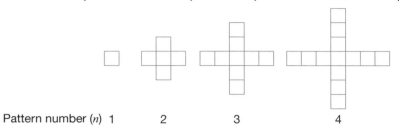

Pattern number (n) 1   2   3   4

Pattern number	Number of squares
1	1 +4
2	5 +4
3	9 +4
4	13

If you want to find the number of squares in pattern number 8 you could draw patterns 5 to 8. You could use a table to record the number of squares used in each pattern instead (see left).

Using the table, the term-to-term rule is 'add 4'.

So the number sequence is 1, 5, 9, 13, 17, 21, 25, 29, ...

Since 29 is the eighth term in this sequence, there are 29 squares in pattern number 8.

In an **arithmetic sequence** (or arithmetic progression) the terms increase or decrease by a fixed number called the **common difference**.

1, 4, 7, 10, ... is an arithmetic sequence with a common difference of 3.

27, 22, 17, 12, ... is an arithmetic sequence with a common difference of -5.

The nth term

When the position-to-term rule of a sequence is given in algebraic form it is called the **nth term**. You need to be able to find the nth term of an arithmetic sequence.

NAILIT!

The nth term is the term at position n in the sequence.

WORKIT!

Write down an expression for the nth term of the following sequence. 5, 9, 13, 17, ...

Term number:	1	2	3	4	...	n
Term:		5	9	13	17	...
Common difference:		+4	+4	+4		+1
4 × term number:		4	8	12	16	...	$4n$

The nth term is $4n + 1$

To work out the nth term:

1. Find the common difference: $+4$

2. Multiply each term number by the common difference:
$1 \times 4 = 4, 2 \times 4 = 8, 3 \times 4 = 12, 4 \times 4 = 16, ... n \times 4 = 4n$

3. Work out how to get from each term in $4n$ to the term in the original sequence: $+1$

You should always check your expression for the nth term by substituting values of n into your expression.

1st term when $n = 1$: $4n + 1 = 4 \times 1 + 1 = 5$ ✓

2nd term when $n = 2$: $4n + 1 = 4 \times 2 + 1 = 9$ ✓

You can also use the nth term to find any term of the sequence.

For the 50th term: $n = 50$, $4n + 1 = 4 \times 50 + 1 = 201$

You may need to decide whether a given number is a term in the sequence, or find the first term that is greater or less than a certain number.

WORKIT!

The nth term of the arithmetic sequence 1, 3, 5, 7, … is $2n - 1$.

a Show that 86 cannot be a term in this sequence.

If 86 is in the sequence, then:

$86 = 2n - 1$ ◄ Write an equation using the nth term. Solve it for n.

Add 1 to both sides. ► $87 = 2n$

$43.5 = n$ ◄ Divide both sides by 2.

The value of n has to be a positive whole number if the number is a term in a sequence. ► But 43.5 is not an integer, so 86 cannot be in the sequence.

b Find the first term in the sequence that is greater than 100.

$100 = 2n - 1$

$101 = 2n$

$50.5 = n$

The 51st term is the first term that is greater than 100:

$2n - 1 = 2 \times 51 - 1 = 101$

DOIT!

The arithmetic sequence 13, 10, 7, 4, 1, … gets smaller and smaller.

Show that the nth term of this sequence is $16 - 3n$.

✓ CHECKIT!

1 a Write down the next **two** terms in the following sequence.

3, 9, 15, 21, …

   b Write down the term-to-term rule of the sequence.

2 The position-to-term rule for a sequence is 'multiply by 4 and subtract 2'.

   a Find the first four terms of this sequence.

   b Find the 20th term of this sequence.

3 Here are the first three patterns in a sequence of patterns made from sticks.

Pattern number 1   Pattern number 2   Pattern number 3

   a Complete the table.

Pattern number	1	2	3	4	5
Number of sticks	7				

   b Find the number of sticks in pattern number 7.

4 The nth term of a sequence is $2n + 3$.

   a Write down the first three terms.

   b Work out the 100th term.

   c Which term has a value of 73?

5 a Here are the first three terms of a sequence.

$-25, -18, -11, …$

Find the first two terms in the sequence that are greater than zero.

   b The nth term of a different sequence is $15 - 2n$.

Which term is the first one with a negative value?

6 a Write down an expression for the nth term of the following sequence.

3, 7, 11, 15, …

   b Show that 100 cannot be a term in this sequence.

Non-linear sequences

SNAPIT! Special sequences to remember

Sequence	
Triangular numbers 1, 3, 6, 10, 15, 21, …	Generated from a pattern of dots that form a triangle:
Square numbers 1, 4, 9, 16, 25, 36, 49, …	The sequence of square numbers: $1^2, 2^2, 3^2, 4^2, 5^2, 6^2,$ … This is also an example of a quadratic sequence.
Cube numbers 1, 8, 27, 64, 125, 216, …	The sequence of cube numbers: $1^3, 2^3, 3^3, 4^4, 5^3, 6^3,$ …
Fibonacci-type 1, 1, 2, 3, 5, 8, …	The next term in the sequence is found by adding the previous two terms together.
Quadratic 2, 5, 10, 17, 26, 37, …	Has n^2 and no higher power of n in its nth term. For the example, the nth term is $n^2 + 1$
Geometric sequence (or geometric progression) 1, 3, 9, 27, 81, 243, …	The next term is found by multiplying the previous term by a fixed number. For the example, the term-to-term rule is 'multiply by 3'.

DOIT!

Triangular numbers

Write down the pattern of common differences for the triangular number sequence 1, 3, 6, 10, 15, 21, …

What can you say about the common difference?

To generate terms in a quadratic sequence (which includes an n^2) you will need to substitute different values of n into the nth term.

WORKIT!

The nth term of a sequence is $2n^2$.
Work out the first two terms and the 10th term of this sequence.

1st term when $n = 1$: $2n^2 = 2 \times 1^2 = 2$
2nd term when $n = 2$: $2n^2 = 2 \times 2^2 = 8$
10th term when $n = 10$: $2n^2 = 2 \times 10^2 = 200$

> Only the value of n is squared: $2n^2$ is not the same as $(2n)^2$.

CHECKIT!

1 Which sequence is a geometric progression?

   a 1, 3, 5, 7, 9, …   **c** 1, 4, 5, 9, 14, …

   b 1, 2, 4, 8, 16, …   **d** 1, 4, 9, 16, 25, …

2 Write down the next **two** terms in the following quadratic sequence.

   3, 6, 11, 18, …

3 Work out the 5th term of the following geometric progression.

   5, 0.5, 0.05, …

4 A Fibonacci-type sequence is made by adding the previous two terms together. The first five terms are 2, 2, 4, 6, 10. Work out the 8th term.

5 Write down the first four terms of a sequence where the nth term is given by $n^2 + 5$.

Show that...

$\frac{10x}{2} \equiv 5x$ is an **identity**. An identity is true for all values of the letters. An identity can be written with the identity symbol '$\equiv$' or with an equals sign.

$2(n + 1) \equiv 2n + 2$ is also an identity. You can show this by expanding the brackets on the left-hand side (LHS) to show that this expression is identical to the one on the right-hand side (RHS).

LHS = $2(n + 1) \equiv 2n + 2$ RHS = $2n + 2$

To show that expressions are equivalent, you need to use algebra – in particular, expanding brackets (page 47) and simplifying expressions (page 43).

SNAPIT!

Symbols

'$=$' means 'is equal to'
    $x + 2 = 6$ when $x = 4$
'$\neq$' means 'is **not** equal to'
    $2 + 2 \neq 5$
'$\equiv$' means 'is identically equal to'
    $2(n + 1) \equiv 2n + 2$

You need to show that the expression on the LHS is equivalent to the expression on the RHS. Expand the brackets and simplify both expressions.

WORKIT!

Show that $(x - 1)^2 - 16 \equiv (x + 3)(x - 5)$

LHS = $(x - 1)^2 - 16 \equiv (x - 1)(x - 1) - 16 = x^2 - x - x + 1 - 16$
$= x^2 - 2x - 15$

RHS = $(x + 3)(x - 5) = x^2 - 5x + 3x - 15$
$= x^2 - 2x - 15$

LHS = RHS
So $(x - 1)^2 - 16 \equiv (x + 3)(x - 5)$

NAILIT!

When you are told to 'Show that ...', write down **every** step of your working.

You also need to have a knowledge of odd and even numbers.

If n is an integer, then $2n$ is even and $2n + 1$ is odd. For example, if $n = 5$, $2n = 2 \times 5 = 10$ (even) and $2n + 1 = 2 \times 5 + 1 = 11$ (odd).

Note that $2n - 1$ is also odd: $2 \times 5 - 1 = 9$.

When you multiply an odd number by an even number, the answer is always even. You could use any of the following examples to test this:
$3 \times 2 = 6, 5 \times 10 = 50, 11 \times 4 = 44$.

STRETCHIT!

Odd or even?
If n is an even number, explain why $(n - 1)(n + 1)$ will always be odd.
Let m be an odd number. Explain why $(m - 1)(m + 1)$ will always be even.

✓ CHECKIT!

1 Show that
    a $4(x - 3) + 2(x + 5) \equiv 3(2x - 1) + 1$ **b** $(x + 2)(x - 2) \equiv x^2 - 4$

Number machines

A **number machine** has an input (x), an output (y) and one or more operation. The operation ($+$, $-$, $\times$, $\div$) tells you what to do to the input to obtain the output.

This is a two-stage number machine as there are two operations to be performed.

In terms of x, $y = 5x - 4$ ('multiply x by 5, then subtract 4').

If you know the output, you can work out the input by working backwards.

In terms of y, $x = \dfrac{y + 4}{5}$

DO IT!

Draw some two-stage number machines on a set of plain cards. Use operations such as $\times 4$, $- 2$.

Shuffle the cards and take one at random. Work out the output (y) for each of the x inputs -2 to $+2$, and an equation for the output y in terms of the input x.

WORKIT!

a Complete the number machine so that $y = 2x + 4$.

$y = 2x + 4$ means 'multiply x by 2, then add 4'.

b Work out the input when the output is 12.

Input $x = (12 - 4) \div 2 = 8 \div 2 = 4$

$-$ is the inverse (opposite) of $+$ and $\div$ is the inverse of $\times$.

c Work out the value of x.

$x = \dfrac{y - 4}{2}$

$\dfrac{y - 4}{2}$ is equivalent to $(y - 4) \div 2$.

Use brackets to show the first operation that must be done. (Without brackets: $12 - 4 \div 2 = 12 - 2 = 10$)

CHECKIT!

1

a Write down the output y in terms of x.

b Work out the output when the input is 10.

c Work out the value of x in terms of y.

d Sasha says, 'There is a value of x whe it is equal to y.'

Show that Sasha is correct.

Coordinates and midpoints

Coordinates

Coordinates are used to describe the position of a point on a coordinate grid. The first number in the coordinate pair (x, y) gives the horizontal position; the second number (x, y) gives the vertical position.

Coordinates can include positive and negative numbers.

WORKIT!

Here is a grid showing points P, Q and R.

a Write down the coordinates of the points P and R.

$P\ (-1, 4)$ $R\ (1, 1)$

b On the grid, mark with a cross the point S so that $PQRS$ is a parallelogram.

Opposite sides of a parallelogram are equal length and parallel. Plot point S so that it is level with R and one square to the left of point P.

Midpoint of a line

The **midpoint** of a line is exactly halfway along the line.

To find the midpoint, add the x coordinates and divide by 2 and add the y coordinates and divide by 2.

Add the y coordinates and divide by 2.

WORKIT!

Find the coordinates of the midpoint of AB.

$A(-3, 0)$, $B(4, 2)$

x coordinate:

$4 + (-3) = 1$ $1 \div 2 = 0.5$

y coordinate:

$2 + 0 = 2$ $2 \div 2 = 1$

Midpoint is $(0.5, 1)$

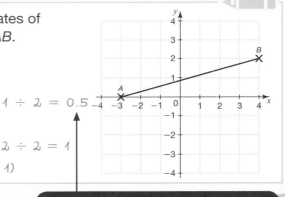

Add the x coordinates and divide by 2.

CHECK**IT!**

1 Here is a grid showing the points A and B.

a Write down the coordinates of the point A.

b On the grid, mark with a cross the point $(-4, -3)$. Label this point C.

c On the grid, mark with a cross a point D so that the quadrilateral $ABCD$ is a rectangle.

d Work out the coordinates of the midpoint of BD.

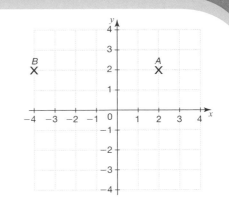

Straight-line graphs

Drawing straight-line graphs of the form $y = mx + c$

SNAPIT! **Horizontal and vertical lines**

A horizontal line on a graph has equation $y = a$, where a is a number.

A vertical line on a graph has equation $x = a$, where a is a number.

The x-axis has equation $y = 0$.

The y-axis has equation $x = 0$.

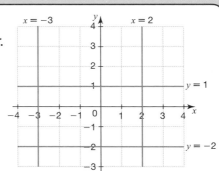

The equation of a straight line on a graph is $y = mx + c$ where m and c are numbers.

m is the **gradient**. This tells you how steep the line is.

Lines that slope upwards from left to right have positive gradients.

Lines that slope downwards from left to right have negative gradients.

c is the **y-intercept**. This is the point where the line crosses the y-axis. The coordinates of this point are $(0, c)$.

SNAPIT! **Graphs of $y = x$ and $y = -x$**

The straight line $y = x$ has gradient 1 and y-intercept $(0, 0)$.

The straight line $y = -x$ has gradient -1 and y-intercept $(0, 0)$.

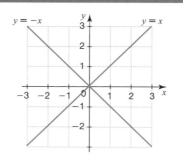

You can compare the equation of a straight line with the general form of the equation $y = mx + c$ to identify the gradient and y-intercept.

The straight line $y = 2x - 4$ has gradient 2 ($m = 2$) and y-intercept $(0, -4)$ ($c = -4$).

The straight line $y = -3x + 1$ has gradient -3 ($m = -3$) and y-intercept $(0, 1)$ ($c = 1$).

To plot a straight-line graph you need to:

1 Make a table of some values of x. Substitute these into the equation to find the values of y.

2 Plot the points from the table on the coordinate grid.

3 Join the points with a straight line.

WORKIT!

Draw the graph of $y = 2x - 1$ for values of x from -1 to 3.

x	-1	0	1	2	3
$y = 2x - 1$	-3	-1	1	3	5

Make a table of values using the values of x given in the question.

Substitute the values of x into $y = 2x - 1$:

When $x = -1$,

$$y = 2 \times -1 - 1$$
$$= -2 - 1 = -3$$

Plot the points on a coordinate grid. The first point is $(-1, -3)$.

Use a ruler to join the points with a straight line.

You can sketch a straight-line graph using the gradient and y-intercept.

WORKIT!

On the grid, draw the graph of $y = -2x + 1$

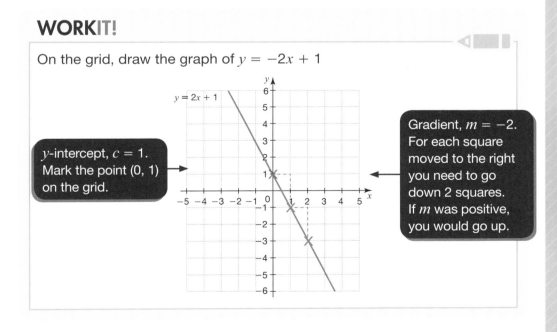

y-intercept, $c = 1$. Mark the point $(0, 1)$ on the grid.

Gradient, $m = -2$. For each square moved to the right you need to go down 2 squares. If m was positive, you would go up.

Finding the equation of a straight line

The **gradient** of a line is a measure of its slope. You can find the gradient of a straight line by drawing a right-angled triangle on the line.

$$\text{Gradient} = \frac{\text{difference in } y \text{ coordinates}}{\text{difference in } x \text{ coordinates}}$$

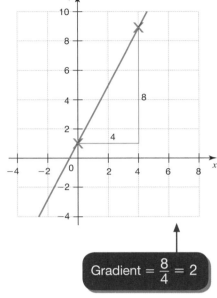

Gradient $= \dfrac{8}{4} = 2$

NAILIT!

To work out the gradient of a straight line:

- Draw a large triangle with a base that is a whole number of units.
- Use the scales on the axes to work out the differences in y and x coordinates.

To write the equation of a line in the form $y = mx + c$ you need to know the gradient (m) and the y-intercept (c).

For the example above, you found that the gradient $m = 2$. The y-intercept can be read off the graph: $c = 1$. So the equation of this line is $y = 2x + 1$.

To find the equation of a straight line through two given points, always draw a sketch to show the two points.

WORKIT!

A has coordinates (1, 2)
B has coordinates (3, 8)
Work out the equation of the straight line AB.

Gradient, $m = \dfrac{8-2}{3-1} = \dfrac{6}{2} = 3$

$y = mx + c$

$m = 3$

So $y = 3x + c$

For point (1, 2), $x = 1, y = 2$:

$2 = 3 \times 1 + c$

$2 = 3 + c$ (−3)

$-1 = c$

Equation of the line AB is $y = 3x - 1$.

Draw a triangle to work out the gradient m.

Both points lie on the line. Use one of the points and substitute the x and y values into the equation.

Write down the general form of the equation of a straight line: $y = mx + c$.
Substitute the value of m into the equation.

Solve the equation for c.

You also need to be able to find the equation of a straight line through one point with a given gradient. Since you are given the gradient m you simply need to work through the last part of the example above to find c.

More straight-line graphs

Lines of the form $ax + by = c$

When the equation of a straight line is given in the form $ax + by = c$, find the x- and y-intercepts to draw its graph.

Use $x = 0$ to find one point:
$2 \times 0 + y = 4$
$y = 4$
Use $y = 0$ to find the second point:
$2x + 0 = 4$
$2x = 4$ $(\div 2)$
$x = 2$

WORKIT!

Draw the graph of $2x + y = 4$.

x	0	2
y	4	0

Plot the two points: (0, 4), (2, 0). Join the points with a straight line.

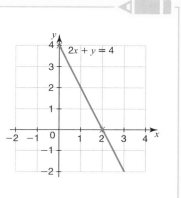

Gradients and parallel lines

The gradient of a straight line is a measure of how steep it is. Steeper lines have larger gradients.

The line $y = 4x - 2$ has a steeper gradient ($m = 4$) than the line $y = 2x + 4$ ($m = 2$).

Parallel lines have the same gradient.

Three of these lines are parallel.
They have a gradient of 2.

$y = 2x$ $y = 2x + 1$ $y = 2x - 1$

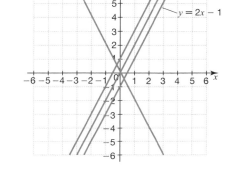

The line $y = -2x$ is not parallel. This line has a gradient of -2.

Parallel lines can be identified by looking at their equations. You may first need to rearrange the equation into the form $y = mx + c$.

If you're unsure about changing the subject of an equation, have a look at page 65.

WORKIT!

Here are the equations of four straight lines.

A $y = 3x + 1$ B $1 - 3x = y$ C $y - 3x = 4$ D $3y = x + 1$

Two of the lines are parallel.

Write down the two parallel lines.

A: $y = 3x + 1$ The equation is already in the form $y = mx + c$

B: $1 - 3x = y$
    $y = -3x + 1$ Rewrite the equation in the form $y = mx + c$

C: $y - 3x = 4$ $(+ 3x)$
    $y = 4 + 3x$
    $y = 3x + 4$ Rearrange the equation into the form $y = mx + c$

D: $3y = x + 1$ $(\div 3)$
    $y = \frac{1}{3}x + \frac{1}{3}$ Rearrange the equation into the form $y = mx + c$

Lines A and C are parallel.   Parallel lines have the same gradient. Lines A and C have gradient 3. Make sure you don't ignore the negative signs. Line B has gradient -3. This means the line slopes downwards from left to right.

y-intercepts

You may also need to rearrange the equation of a straight line into the form $y = mx + c$ to identify the *y*-intercept (c).

For example:

$$3y = x + 6$$

> Divide both sides by 3.

$$y = \frac{x}{3} + 2$$

The *y*-intercept is (0, 2).

Using graphs to solve linear equations

> Solving linear equations algebraically is covered on page 55.

You can use graphs to solve linear equations such as $2x + 1 = 0$ and $3x - 2 = 3$. This is a **graphical method**.

You solve linear equations such as $2x + 1 = 0$ by reading off the x coordinate where the graph $y = 2x + 1$ crosses the x-axis ($y = 0$).

WORKIT!

a Complete the table of values for $y = 2x + 1$

> Substitute the values of x into $y = 2x + 1$
>
> When $x = -2$,
> $y = 2 \times (-2) + 1$
> $= -4 + 1 = -3$

x	−2	−1	0	1	2
y	−3	−1	1	3	5

b Draw the graph of $y = 2x + 1$.

c Use your graph to find an estimate for the value of y when $x = 1.4$.

When $x = 1.4$, $y = 3.8$

> Read off the values from the graph.

d Use your graph to find an estimate for the value of x when $y = 2.8$.

When $y = 2.8$, $x = 0.9$

e Use your graph to write down the solution of $2x + 1 = 0$.

$x = -0.5$

> Find where the graph crosses the x-axis. This is where $y = 0$. Read off the x value.

You solve linear equations such as $3x - 2 = 3$ by reading off the x coordinate at the point of intersection of the graph $y = 3x - 2$ with the graph $y = 3$.

STRETCHIT!

Use the graph of $y = 3x - 2$ on the next page to solve the equation $3x - 2 = -x$.

WORKIT!

Here is the graph of
$y = 3x - 2$.

Use the graph to solve
the equation $3x - 2 = 3$.

$x \approx 1.7$

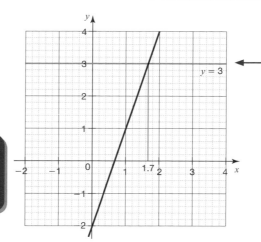

Read off the x value as
accurately as possible.
The $\approx$ sign means
'approximately equal to'.

To solve the equation $3x - 2 = 3$
you need to find the point of
intersection of the graphs
$y = 3x - 2$ and $y = 3$.
You are given the graph of
$y = 3x - 2$.
Draw the line $y = 3$ on the graph.
This is a horizontal line parallel to
the x-axis.

✓ CHECKIT!

1 Draw the graph of $y = -x + 2$ for
values of x from -2 to 2.

2 Draw the graph of $y = 2x - 3$.

3

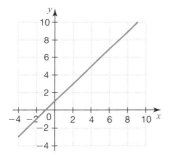

Find the equation of the straight line.

4 A straight line has gradient 2.

The point $(1, -2)$ lies on the line.

Find the equation of the line.

5 A has coordinates $(0, 4)$.

B has coordinates $(4, 2)$.

Work out the equation of the straight line AB.

One of the points is $(0, 4)$ so you
can write down the y-intercept
of the line without calculation.

6 Draw the graphs of

a $x + y = 6$ **b** $2y + x = 4$

7 Show that the lines $4x + 4y = 8$ and
$2x = 6 - 2y$ are parallel.

8 $y + 2x = 4$

Write down the gradient and the
coordinates of the y-intercept.

9 Here is the graph of $y = 2x - 1$.

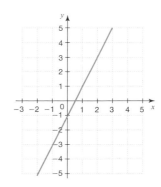

a Use the graph to find an estimate
for y when $x = 2.2$.

b Use the graph to write down the
solution of $2x - 1 = 0$

c Use the graph to write down the
solution of $2x - 1 = 4$

d Use the graph to write down the
solution of $2x - 1 = -2$

Solving simultaneous equations

Algebraic solution

When there are two unknowns you need two equations. These are called **simultaneous equations**. You need to find the values for the two unknowns which make **both** equations true.

Simultaneous equations can be solved using the substitution or elimination method. For both methods you need to be able to solve linear equations.

Substitution method

WORKIT!

Solve the simultaneous equations

$$y = x + 4 \quad (1)$$
$$3x + y = 8 \quad (2)$$

Number each equation.

Substitute equation (1) into equation (2): $3x + x + 4 = 8$

Collect like terms. $4x + 4 = 8$

Subtract 4 from both sides. $4x = 4$

Divide both sides by 4. $x = 1$

Substitute $x = 1$ into equation (1): $y = 1 + 4$

$y = 5$

Solution is $x = 1, y = 5$

Check: $3x + y = 3 \times 1 + 5 = 3 + 5 = 8$ ✔

NAIL IT!

Check your answer by substituting both values into the original equation that you haven't used.

Elimination method

WORKIT!

Solve the simultaneous equations

$2x + y = 5 \quad (1)$

$x - 3y = 6 \quad (2)$

$3 \times (1): 6x + 3y = 15 \ (3)$

$(3) + (2): 7x = 21 \quad (\div 7)$

$x = 3$

Substitute $x = 3$ into (1):

$2 \times 3 + y = 5$

$6 + y = 5 \quad (- 6)$

$y = -1$

Solution is $x = 3, y = -1$

Check: $x - 3y = 3 - 3(-1) = 3 + 3 = 6$ ✔

The signs of the terms in y are different so choose this unknown to eliminate.

Multiply equation (1) by 3 to make the coefficients of y (the numbers in front of the ys) equal.

NAIL IT!

When deciding which unknown to eliminate, choose the unknown where the signs are different if possible. Then you can add the equations to eliminate the unknown.

For these simultaneous equations, both equations must be multiplied by a number before an unknown can be eliminated.

$2x + 4y = 10$ (1) ← The signs of the terms in both x and y are the same in both equations. Choose to eliminate x.

$3x + 3y = 12$ (2)

$6x + 12y = 30$ (3) ← Multiply (1) by 3 and (2) by 2 to make the coefficients of x equal.

$6x + 6y = 24$ (4)

$6y = 6$ ← Subtract (4) from (3) to eliminate the terms in x.

$y = 1$ ← Divide both sides by 6.

$2x + 4 = 10$ ← Substitute $y = 1$ into (1) to find the value of x.

$2x = 6$ ← Subtract 4 from both sides.

$x = 3$ ← Divide both sides by 2.

Solution is $x = 3$, $y = 1$.

DO IT!

Sometimes the coefficients of the terms in x or the terms in y are equal so you do not need to multiply before you can eliminate an unknown.

Solve the simultaneous equations

$2x + y = 7$ (1)

$x - y = 5$ (2)

by adding (1) and (2) to eliminate the terms in y.

Setting up simultaneous equations

You can set up simultaneous equations to represent real-life situations and solve problems.

You need to define the unknowns, then turn each sentence in the problem into an equation.

WORKIT!

Two coffees and a muffin cost £3.75.

Three coffees and two muffins cost £6.00.

Work out the cost of a coffee and the cost of a muffin.

Let c = coffee and m = muffin. ← Define the unknowns. You can use any letters to represent them.

$\qquad 2c + m = 375$ (1)

$\qquad 3c + 2m = 600$ (2) ← Turn each sentence into an equation. It is easier to work in pence.

$2 \times$ (1): $4c + 2m = 750$ (3)

$\qquad$ (3) $-$ (2): $c = 150$ ← Once the equations have been set up, solve the simultaneous equations as usual.

Substitute $c = 150$ into (1): $2 \times 150 + m = 375$

$\qquad 300 + m = 375$

$\qquad m = 75$ ← Subtract 300 from both sides.

Check: $3c + 2m = 3 \times 150 + 2 \times 75 = 450 + 150 = 600$ ✓

A coffee costs £1.50.

A muffin costs 75p. ← Make sure you answer the question asked by stating the cost of a coffee and the cost of a muffin.

NAIL IT!

When setting up simultaneous equations to represent a real-life situation, make sure that both sides of each equation use the same units.

Graphical solution

You can solve simultaneous equations by drawing the graphs of the two equations and finding the point of intersection. If you're unsure about drawing straight-line graphs, have a look at page 74.

WORKIT!

Solve the simultaneous equations

$x + 2y = 4$

$x - y = 1$

For each equation, plot the two points and join with a straight line.

Find the x-intercept ($y = 0$) and y-intercept ($x = 0$) of both equations.

$x + 2y = 4$

x	0	4
y	2	0

$x - y = 1$

x	0	1
y	-1	0

Solution is $x = 2$, $y = 1$

Check: $x - y = 2 - 1$
$= 1$ ✔

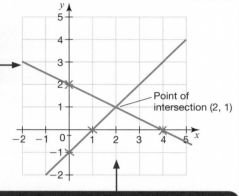

Point of intersection (2, 1)

Find the coordinates of the point where the two lines cross. This gives the solution to the simultaneous equations.

Check your answer by substituting into either of the original equations.

CHECKIT!

1 $x + y = 16$
$x - y = 5$
Work out the values of x and y.

2 Solve these simultaneous equations
You **must** show your working.

a $2x + y = 4$
$3x - y = 1$

b $x - y = 5$
$2x + y = 4$

c $2x + y = 8$
$x + y = 2$

d $4x - y = 10$
$x + 2y = 7$

e $2x + y = 7$
$x - 4y = 8$

f $2x + 3y = 7$
$3x - 2y = 4$

3 The sum of two numbers is 21.
The difference between the numbers is 7.
Work out the two numbers.

Let the two numbers be x and y.

4 3 burgers and 2 colas cost £5.05.
3 burgers and 4 colas cost £7.25.
Find the cost of a burger and the cost of a cola.

5 Use a graphical method to solve these simultaneous equations.

a $x + y = 10$
$y - 3x = 2$

b $x - y = 7$
$x + 4y = 12$

Quadratic graphs

A **quadratic function** contains an x^2 term but no higher power of x.

$y = x^2$, $y = 2x^2 + 1$, $y = 5 - x^2$ and $y = 4x^2 - 2x + 3$ are all examples of quadratic functions.

The graph of a quadratic function is a curved shape. A quadratic function with a positive x^2 term has a ∪-shaped curve. A quadratic function with a negative x^2 term has an upside-down ∪-shaped curve.

A function is a rule you apply to a set of values, giving another set of values. In an equation like $y = x^2 - 1$, y is a function of x.

SNAPIT!   Graphs of $y = x^2$ and $y = -x^2$

Minimum point

Maximum point

$y = x^2$

$y = -x^2$

NAILIT!

Drawing quadratic graphs

- Plot each point with a cross (×).
- Draw a smooth curve through the points.

The curve should pass through **all** the points!

The minimum or maximum point is called a **turning point** as this is where the graph turns.

All quadratic graphs have a **line of symmetry**.

To draw a quadratic graph you need to complete a table of values.

WORKIT!

a Complete the table of values for $y = x^2 + 2$.

x	−3	−2	−1	0	1	2	3
y	11	6	3	2	3	6	11

b Draw the graph of $y = x^2 + 2$.

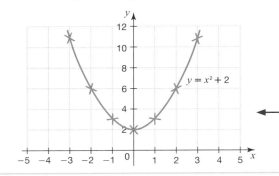

$y = x^2 + 2$

Substitute each value of x into $x^2 + 2$ to work out the corresponding value of y.

Put negative values of x in brackets and use the correct order of operations (**BIDMAS**).

When $x = -3$, $y = (-3)^2 + 2$
$= 9 + 2 = 11$

When $x = 2$, $y = 2^2 + 2$
$= 4 + 2 = 6$

Plot the points from your table on the graph: $(-3, 11)$, $(-2, 6)$…

Join your points with a **smooth** curve.

WORKIT!

a Draw the graph of $y = x^2 + 2x - 3$ for values of x from -4 to 2.

x	-4	-3	-2	-1	0	1	2
y	5	0	-3	-4	-3	0	5

> Take care with the negative values.
> When $x = -4$, $y = (-4)^2 + 2(-4) - 3$
> $= 16 - 8 - 3$
> $= 5$

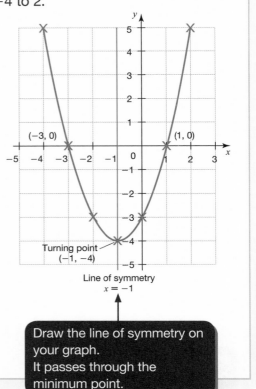

b Write down the equation of the line of symmetry.

The line of symmetry has equation $x = -1$

c Write down the coordinates of the turning point.

$(-1, -4)$

> The turning point is the **minimum point**. Read off the coordinates of this point from your graph.

> Draw the line of symmetry on your graph.
> It passes through the minimum point.

Using graphs to solve quadratic equations

A **quadratic equation** contains an x^2 term but no higher power of x.
$x^2 - x - 4 = 0$ and $x^2 - 3x - 2 = -3$ are both examples of quadratic equations.

You can find the solutions (or **roots**) to a quadratic equation by drawing a graph.

You solve quadratic equations such as $x^2 - x - 4 = 0$ by reading off the x coordinates where the graph $y = x^2 - x - 4$ crosses the x-axis ($y = 0$).

DOIT!

Copy and complete this table of values for $y = x^2 + 2x - 3$ for values of x from -4 to 2.

x	
x^2	
$+2x$	
-3	
y	

Compare this table with the one above in Work it! Which do you prefer?

WORKIT!

Here is the graph of $y = x^2 - x - 4$.

Use the graph to solve the equation

$x^2 - x - 4 = 0$

$x = 2.6$ and $x = -1.6$

> Find where the graph crosses the x-axis. This is where $y = 0$.
> Read off the values of x.
> Make sure you write down **both** solutions.

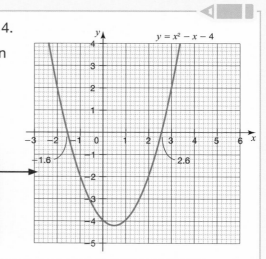

You solve quadratic equations such as $x^2 - 3x - 2 = 3$ by reading off the x coordinate at the points of intersection of the graph $y = x^2 - 3x - 2$ with the graph $y = 3$.

WORKIT!

Here is the graph of $y = x^2 - 3x - 2$.

Use the graph to write down the solutions to $x^2 - 3x - 2 = -3$.

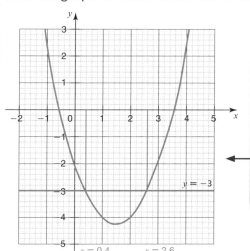

To solve the equation $x^2 - 3x - 2 = -3$ you need to find the points of intersection of the graphs:
$y = x^2 - 3x - 2$
$y = -3$
You are given the graph of
$y = x^2 - 3x - 2$
Draw the line $y = -3$ on the graph. This is a horizontal line parallel to the x-axis.

$x = 0.4$ and $x = 2.6$

Read off the x value of the points where the line crosses the curve.

DOIT!

Draw the graph of $y = x^2$ for values of x between -3 and 3. Use your graph to find the approximate values of x when $y = 3$.

STRETCHIT!

Use the graph of
$y = x^2 - 3x - 2$ to solve the equation
$x^2 - 3x = 3$

Rearrange the equation into the form
$x^2 - 3x - 2 = \bigcirc$

CHECKIT!

1 a Draw the graph of $y = x^2 - 2x - 2$ for values of x from -2 to 4.

  b Write down the equation of the line of symmetry.

  c Write down the coordinates of the turning point.

2 a Draw the graph of $y = x^2 - 2x$ for values of x from -2 to 4.

  b Use your graph to write down the roots of $y = x^2 - 2x$.

3 a Draw the graph of $y = x^2 + 2x - 1$ for values of x from -3 to 3.

  b Using your graph, write down the roots of $x^2 + 2x - 1 = 0$.

Solving quadratic equations

A **quadratic equation** contains an x^2 term but no higher power of x.
$x^2 - 9 = 0$, $x^2 + 2x = 0$, $x^2 + 4x + 4 = 0$ and $x^2 - 3x - 2 = 0$ are all examples of quadratic equations.

You can solve quadratic equations by factorising.

> If you're unsure about factorising, have a look at page 49.

$$x^2 + 2x = 0$$
$$x(x + 2) = 0$$

Factorise

Either $x = 0$ or $x + 2 = 0$.

> Solve the linear equation $x + 2 = 0$

So $x = 0$ or $x = -2$.

To solve an equation of the form $x^2 - 9 = 0$ use the difference of two squares:
$x^2 - a^2 = (x + a)(x - a)$.

> $x^2 - 9 = x^2 - 3^2$

$$x^2 - 9 = 0$$
$$(x + 3)(x - 3) = 0$$

Either $x + 3 = 0$ or $x - 3 = 0$.

> Solve the linear equations $x + 3 = 0$ and $x - 3 = 0$

So $x = -3$ or $x = 3$.

DOIT!

Solve the equation $x^2 - 9 = 0$ using this alternative method.
1 Rearrange to make x^2 the subject.
2 Take the square root of both sides.
3 Give the negative and positive solutions.

WORKIT!

Solve $x^2 + 4x + 4 = 0$

$$x^2 + 4x + 4 = 0$$
$$(x + 2)(x + 2) = 0$$

Factorise. The factor pairs of 4 are 1×4 and 2×2. The pair that add up to 4 are 2 and 2.

Either $x + 2 = 0$ or $x + 2 = 0$

So $x = -2$

Solve the linear equation $x + 2 = 0$
Note that here there is only **one** solution.

Check: $(-2)^2 + 4 \times (-2) + 4 = 4 - 8 + 4 = 0$ ✓

STRETCHIT!

Solve these quadratic equations
$$\frac{x^2}{2} = 8$$
$$2x^2 = 50$$

$y = x^2 - 4x$ and $y = x^2 - 49$ are both examples of quadratic functions. The roots of a quadratic function are the solutions when $y = 0$.

$$y = x^2 - 4x$$

> Set $y = 0$.

$$x^2 - 4x = 0$$
$$x(x - 4) = 0$$

Factorise.

Either $x = 0$ or $x - 4 = 0$.

> Solve the linear equation $x - 4 = 0$.

So the roots of the function are $x = 0$ or $x = 4$.

CHECKIT!

1 Solve

  a $x^2 - 4x = 0$

  b $x^2 + 7x = 0$

  c $x^2 - 16 = 0$

  d $x^2 + 10x + 9 = 0$

  e $x^2 + x - 12 = 0$

  f $x^2 - 6x - 16 = 0$

2 Find the roots of these functions.

  a $y = x^2 - 49$

  b $y = x^2 - 3x$

  c $y = x^2 + 7x + 6$

Cubic and reciprocal graphs

Cubic graphs

A **cubic function** has an x^3 term and no higher power of x.

$y = x^3$, $y = x^3 + 1$ and $y = x^3 - x$ are all examples of cubic functions.

> ## SNAP IT!   General shape of a cubic graph
>
>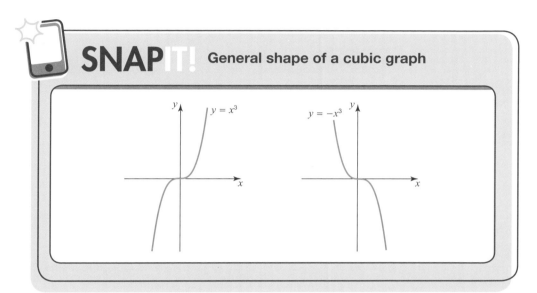

To draw a cubic graph you need to complete a table of values.

WORKIT!

a Complete the table of values for $y = x^3 + 2$

x	−2	−1	0	1
y	−6	1	2	3

> Substitute each value of x into $y = x^3 + 2$ to work out the corresponding value of y.
>
> Take care with negative numbers:
> $(-1)^3 = -1 \times -1 \times -1 = -1$
>
> When $x = -2$, $y = (-2)^3 + 2$
> $= -8 + 2 = -6$

b Draw the graph of $y = x^3 + 2$

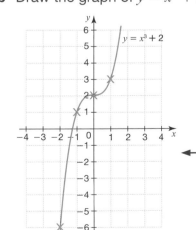

> Plot the points from your table on the graph: $(-2, -6)$...
>
> Draw a smooth curve through **all** the points.

Reciprocal graphs

The **reciprocal** of x is $\frac{1}{x}$

$y = \frac{1}{x}$ is an example of a **reciprocal function**.

The graph of a reciprocal function has two parts. It does not cross or touch the x-axis or y-axis.

This is the graph of $y = \frac{1}{x}$

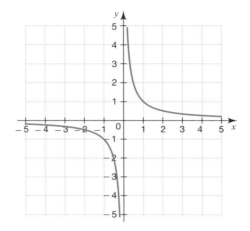

DOIT!

Compare the graph of $y = \frac{1}{x}$ to the graph of $y = -\frac{1}{x}$ (see Check it!). How does the negative sign affect the shape of the graph?

CHECKIT!

1 a Complete the table of values for $y = x^3 - 1$

x	−3	−2	−1	0	1	2	3
y		−9					26

b Draw the graph of $y = x^3 - 1$

2 a Complete the table of values for $y = -\frac{1}{x}$

x	−4	−3	−2	−1	$-\frac{1}{2}$	$-\frac{1}{4}$	$\frac{1}{4}$	$\frac{1}{2}$	1	2	3	4
y	$\frac{1}{4}$	$\frac{1}{3}$			2		−4			$-\frac{1}{2}$		

b Draw the graph of $y = -\frac{1}{x}$

Drawing and interpreting real-life graphs

Linear graphs

Conversion graphs can be used to convert one unit into another. They can also be used to convert between different currencies.

Straight-line graphs used to represent a real-life situation such as a fuel bill will not go through the origin (0, 0) if there is a fixed or standing charge.

DO IT!

Use the information in the table to draw a conversion graph for temperatures in degrees Fahrenheit (°F) and degrees Celsius (°C).

°F	−22	32	140
°C	−30	0	60

Use the graph to convert 40 °C to °F.

WORKIT!

The graph shows the charge for electricity, in £s.

The cost consists of a standing charge plus a charge per unit of electricity used.

a Write down the standing charge.

   Standing charge = £20

b Work out the cost per unit of electricity.

   Gradient = $\frac{30}{120}$ = 0.25

   Cost per unit of electricity is 25p.

> Read off the cost of 0 units from the graph.

> The gradient
> $$\frac{\text{cost (£)}}{\text{number of units used}}$$
> of the graph gives the cost per unit of electricity.
>
> Draw a large right-angled triangle on the line. Make sure your base is a whole number of units.
>
> Use the scales on the axes to work out the differences in the x and y coordinates.

For some questions you cannot simply read off a value directly from the graph.

WORKIT!

This graph can be used to convert between kilometres and miles.

Use the graph to convert 80 km into miles.

From the graph, 40 km = 25 miles

So 80 km = 50 miles

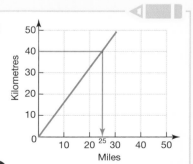

> The graph does not go up to 80 km.
> Read off the number of miles in 40 km and multiply by 2.

> You could read off the value for 10 km and multiply by 8, or read off the value for 20 km and multiply by 4. However, the value for 40 km will give greater accuracy.

Distance–time graphs

A **distance–time graph** represents a journey. It is an example of a **rate of change** graph, as it shows how distance changes over time.

You can write down information about the journey by looking at the shape of the graph.

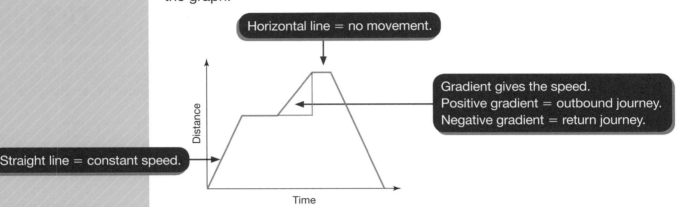

Horizontal line = no movement.

Gradient gives the speed.
Positive gradient = outbound journey.
Negative gradient = return journey.

Straight line = constant speed.

On a distance–time graph the vertical axis represents the distance from the starting point. The horizontal axis represents time. The time could represent 'time taken' given in hours, minutes or seconds, or it could represent the 'time of day' given as, for example, 09:00, 14:00 (using the 24–hour clock).

WORKIT!

Bradley went for a bike ride.

He left home at 7 am.

The distance–time graph represents his bike ride.

At 9 am Bradley stopped for a rest.

a For how many minutes did he rest at 9 am?

   30 minutes

You are told that Bradley stops for a rest at 9 am. The horizontal line represents no movement. He rested between 9 am and 9.30 am so 30 minutes.

b How far was Bradley from home at 10.30 am?

   22 km

c How far did Bradley ride in total?

   44 km

The question is asking for the total distance: outbound journey + return journey. Multiply the outbound journey by 2.

d Find Bradley's average speed on his journey home.

   Average speed = $\frac{22}{2.5}$ = 8.8 km/h

The gradient of the graph gives you Bradley's speed. This represents $\frac{distance}{time}$ = speed.
You need to find the speed of the return journey. This is represented by the line between 10.30 am and 1 pm.
Give the units with your answer: $\frac{distance\ (km)}{time\ (hours)}$ = speed (km/hr)

Speed–time graphs

A **speed–time graph** also represents a journey. It shows how speed changes over time.

You can write down information about the journey by looking at the shape of the graph.

On a speed–time graph the vertical axis represents speed. The horizontal axis represents time.

> Horizontal line = constant speed (no acceleration).

> Gradient gives the acceleration.
> Positive gradient = acceleration.
> Negative gradient = deceleration.

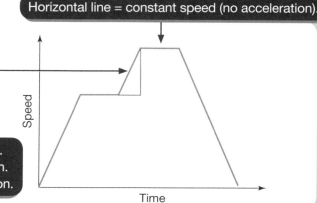

> The gradient of the graph gives you the train's acceleration. This represents
> $$\frac{\text{change in speed}}{\text{time}} = \text{acceleration.}$$
> The first part of the journey is between 0 and 40 seconds. Give the units with your answer:
> $$\frac{\text{change in speed (m/s)}}{\text{time (s)}} =$$
> acceleration (m/s²)

> The acceleration is negative. This means that the train is slowing down (decelerating).

WORKIT!

The graph shows information about the journey of a train between two stations.

a What is the train's maximum speed?

   70 m/s ◄ Read off the value from the graph.

b Describe the train's journey between 40 and 120 seconds.

   The train is travelling at a constant speed of 70 m/s. ◄

> A horizontal line represents constant speed. The acceleration is zero.

c Work out the train's acceleration for the first part of the journey.

   Acceleration = $\frac{70}{40}$ = 1.75 m/s²

d Work out the train's acceleration for the final part of the journey.

   Acceleration = $-\frac{70}{80}$ = −0.875 m/s²

Interpreting graphs

For a container being filled with water, a graph can be drawn to show how the depth of water changes over time. This is also an example of a rate of change graph.

For a container with straight vertical sides, the graph will be a straight line as it fills at a constant rate. A narrow container will fill faster than a wider container. On a graph this would be represented by a steeper line.

A container with a slanting side will produce a curved graph. For a container with a narrower bottom, the water level rises quickly at first but then slows down as the container widens.

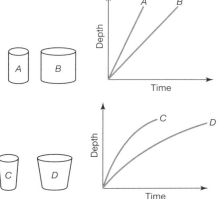

DOIT!

Water is poured into these containers.

Sketch a graph to show depth against time for each container.

CHECKIT!

1 You can use the information in the table to convert between pounds (£) and euros (€).

Pounds (£)	0	20	50	80
Euros (€)	0	24	60	96

a Use the information in the table to draw a conversion graph.

b Use the graph to convert

    i £30 to €

    ii €90 to £

Hannah wants to buy a ring.

The ring costs €100 in France.

The same ring costs £90 in the United Kingdom.

c In which country is the ring cheaper? Give a reason for your answer.

2 This graph shows the cost of using a mobile phone for one month.

The cost includes a fixed monthly charge and charge per minute of calls above the limit of the contract.

a Write down the monthly charge.

b Work out the charge per minute of calls.

3 The cost of hiring a floor sander is given by the formula $C = 12d + 20$, where d is the number of days for which the sander is hired and C (£) is the total cost of hire.

a Draw the graph of $C = 12d + 20$.

b What is the deposit required for hiring the sander?

Another shop hires out floor sanders where the cost of hire is given by the formula $C = 10d + 30$.

c Which shop would you use to hire the sander for more than 5 days? You **must** show your workings.

4 Elinor drives from her house to her friend's house.

The graph represents part of her journey.

One hour after setting off she stopped for a break.

a For how many minutes did she rest?

Elinor arrived at her friend's house $1\frac{1}{2}$ hours after her break.

b How far is Elinor's house from her friend's house?

c Work out Elinor's speed before and after her break.

Elinor stayed at her friend's house for 2 hours. She then drove home at a steady speed.

It took her 2 hours to get home.

d Complete the graph.

5 A ball is thrown vertically upwards.

The graph shows the path followed by the ball.

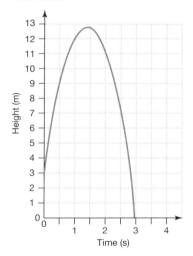

a What is the maximum height reached by the ball?

b How long was the ball in the air?

c There were two times when the ball was at a height of 8 m.

Write down these two times.

d Give a practical reason as to why the graph starts at 3 m.

6 Part of a cyclist's training session is described below.

A cyclist sets off from a resting position and accelerates to a speed of 10 m/s in 20 s.

She then cycles at a constant speed of 10 m/s for a further 50 s.

She then accelerates to a speed of 20 m/s in 30 s.

a Complete the speed–time graph to show the cyclist's session.

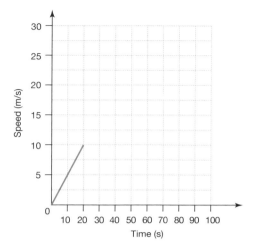

b Work out the cyclist's acceleration over the final 30 s.

7 The graph shows the depth of water (cm) in a bath.

Between O and A the hot and cold water taps are on.

Between B and C a person got into the bath.

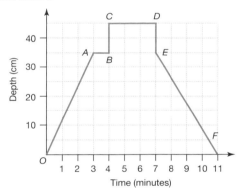

a What was the maximum depth of water before the person got in?

b What was happening between C and D?

c What was happening between D and E?

d Which was quicker, running water into the bath or emptying it? Give a reason for your answer.

Algebra

 You may **not** use a calculator for these questions.

1

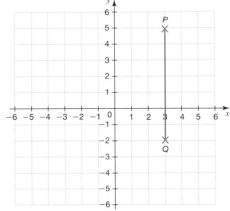

a Write down the coordinates of the point P.

b On the grid mark with a cross (x) the point $(-4, -2)$. Label this point R.

c On the grid, mark with a cross (x) a point S so that the quadrilateral $PQRS$ is a square.

d Work out the coordinates of the midpoint of QS.

e On the grid, draw the line with equation $y = -4$.

2 a Solve $2x + 8 = 4$.

b Here are three expressions.

A $\dfrac{y}{x}$ B $x - y$ C xy

When $x = 2$ and $y = -4$, which expression has the smallest value?

c George says, 'When $x = 4$, the value of $3x^2$ is 144.'

Millie says 'When $x = 4$, the value of $3x^2$ is 48.'

Who is correct? Give a reason for your answer.

3 a Simplify $7a - (3a + 4)$.

b Factorise $8x + 12$.

c Simplify $m^4 \times m$.   **d** Simplify $\dfrac{x^8}{x^3}$.

4 a Complete the table of values for $y = 2x + 1$.

x	-1	0	1	2	3
y			3		

b Draw the graph of $y = 2x + 1$.

c Write down the gradient of the line.

5 Solve $4x + 4 = x + 13$.

6 a Write down the inequality shown on the number line.

b Show the inequality $-1 < x \le 3$ on a number line.

c Write down all the integers that satisfy $1 \le x < 4$.

d Solve

$4x + 2 \le 2x + 5$

7 Luke is x years old.

His grandfather is 65 years older.

His grandfather is also 6 times as old as Luke.

a Write an expression, in terms of x, for his grandfather's age.

b Form an equation in x to work out Luke's age.

8 Here are the first four terms of a sequence.

3, 9, 15, 21

a Write down the next **two** terms of this sequence.

b Can 44 be a term in this sequence? Give a reason for your answer.

The nth term of a different sequence is $2n^2 - 3$.

c Work out the 5th term of this sequence.

9 Expand and simplify $(x + 3)(x + 4)$.

10 a and b are whole numbers.

$a > 30$ $b < 20$

Work out the smallest possible value of $a - b$.

11 Here is a rectangle.

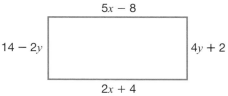

Work out the value of x and the value of y.

12 Work out the values of a and b in the identity

$3(ax - 4) + 2(4x + b) \equiv 14x - 6$.

13 Simplify

  a $9m - 4m + 7m$ **c** $12x \div 2$

  b $3p \times 4p$

14 a Solve $5(w - 4) = 35$.

  b Work out the value of $5a + 7b$ when $a = 7$ and $b = -2$.

  c Write an expression for the total cost of five cakes at a pence each and eight biscuits at b pence each.

15 Three vertices (corners) of a rectangle $PQRS$ are P (3, 4), Q (3, −2) and R (−5, −2).

  a Work out the coordinates of vertex S.

  b Write down the length and width of the rectangle.

16 You can use the information in the table to convert between inches and centimetres.

Inches	0	2	8	12
Centimetres	0	5	20	30

  a Use the information in the table to draw a conversion graph with axes going up to 60 centimetres and 26 inches.

  b Use the graph to convert

   i 10 inches to centimetres

   ii 50 centimetres to inches.

  Arwel needs 60 inches of wooden beading.

  The beading costs 2p per centimetre.

  c Work out the cost of the beading Arwel needs.

17 This rule can be used to work out the cost, in pounds, of hiring a fancy dress costume.

  Multiply the number of days by 12.50 and then add 10.

  Richard hires a costume for x days.

  The total cost is £T.

  a Write down a formula for T in terms of x.

  Suzanne hires a costume.

  The total cost is £72.50.

  b For how many days did Suzanne hire the costume?

18 Make Q the subject of the formula $P = \frac{Q}{4} + R$.

19 Factorise

  a $m^2 + 8m$

  b $x^2 + 7x + 12$

20 a Write down an expression for the nth term of the following sequence.

   2, 5, 8, 11, 14, …

  b Work out the 50th term of the sequence.

  c The nth term of a different sequence is $2n - 3$. Explain why 112 cannot be a term in this sequence.

21 a Expand and simplify $4(x + 5) - 3(2x - 1)$.

  b Simplify $4a^3b^2 \times 5a^2b$.

22 This quadrilateral has dimensions shown.

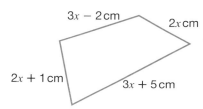

  The perimeter of the quadrilateral is 49 cm. Work out the value of x.

23 Here are two number machines, A and B.

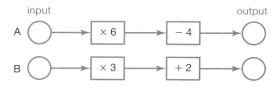

  Both machines have the same input.

  Work out the input that makes the output of A four times the output of B.

24 An arithmetic progression starts $4 + 2a$, $4 + 4a$, $4 + 6a$.

  If the fifth term is 64, work out the value of a.

25 Which equation has roots 2 and −4?

  A $x^2 - 8 = 0$

  B $(x + 2)(x - 4) = 0$

  C $(x - 2)(x + 4) = 0$

  D $2x(x - 4) = 0$

Ratio, proportion and rates of change

Units of measure

You need to be able to convert between units of measure.

SNAP IT! **Converting between units of measure**

Time:	1 year	= 365 days (366 days in a leap year)	Length:	1 km	= 1000 m
				1 m	= 100 cm
				1 cm	= 10 mm
	1 year	= 12 months	Area:	1 m²	= 10 000 cm²
	1 day	= 24 hours		1 cm²	= 100 mm²
	1 hour	= 60 minutes	Volume:	1 litre	= 1000 ml
				1 ml	= 1 cm³
	1 minute	= 60 seconds	Weight:	1 kg	= 1000 g

If you know these standard conversions, you can multiply or divide to convert between units.

NAIL IT!

Multiply to convert to a smaller unit of measure.

Divide to convert to a larger unit of measure.

Use your common sense – are you expecting the answer to be smaller or larger after conversion?

WORKIT!

1   Convert 2.4 km into m.

$2.4 \times 1000 = 2400$ m

> 1 km = 1000 m
> Since you are moving to a smaller unit of measure, you must multiply.

2   Convert 720 g into kg.

$720 \div 1000 = 0.72$ kg

> 1 kg = 1000 g
> Since you are moving to a larger unit of measure, you must divide.

WORKIT!

1   How many minutes are there in half a day?

Half a day = 12 hours

$12 \times 60 = 720$ minutes

> Multiply, since the unit is smaller.

2   Work out the perimeter of this rectangle.

$125 \div 10 = 12.5$ cm

> Divide by 10 since the unit is larger.

Perimeter = $12.5 + 12.5 + 3.2 + 3.2$

= 31.4 cm

> Make sure you give the correct unit in the answer.
> You could convert all the units to mm instead of cm. If it doesn't say in the question which units to use, you can use either.

125 mm

3.2 cm

Sometimes you will be given the conversion to use. Just use the same rule – multiply if you are moving to a smaller unit of measure, divide if you are moving to a larger unit of measure.

WORKIT!

A length of wood measures 17 inches. What is its approximate length in centimetres?

1 inch ≈ 2.5 cm

17 × 2.5 ≈ 42.5 cm

Multiply since the units are smaller.

NAILIT!

≈ means 'approximately equal to' and is used when the conversion isn't exact.

DOIT!

Work out

a how many seconds you spent doing maths work today

b how long your arm is in metres

c how much you weigh in grams

d the area in m² of a square with sides measuring 10 cm by 10 cm.

CHECKIT!

1 Convert these measurements to the units shown in brackets:

a 3 km (m)

b $1\frac{1}{4}$ hours (mins)

c 1.3 m² (cm²)

d 3520 ml (l)

e 2 hours (seconds)

f 14 000 g (kg)

2 A clothed baby weighs 4.5 kg. If his clothes weigh 325 g, what is the actual weight of the baby?

3 A bag weighs 5 pounds. What is the weight in kg?

1 kg ≈ 2.2 pounds

Ratio

A ratio is a way of expressing the relationship between two different quantities.

For example, the ratio of the number of girls to the number of boys in a class is **5 : 4**.

This tells you that for every **5** girls there are **4** boys.

If there are 10 girls (2 × **5**) we know there will be 8 boys (2 × **4**).

You can cancel a ratio down by dividing all the numbers by the same amount. You can find equivalent ratios by multiplying all the numbers by the same amount.

For example, the ratio 5 : 4 is equivalent to the ratio 10 : 8 or 15 : 12 or 2.5 : 2.

A ratio can be reduced to its simplest form by dividing all the parts by the highest common factor. ← For more on the highest common factor, see page 10.

For more on the highest common factor, see page 10.

WORKIT!

Simplify the ratio 4 : 8 : 32.

1 : 2 : 8 ← The highest common factor of 4, 8 and 32 is 4, so divide each part of the ratio by 4.

It may be useful to write a ratio in the form 1 : n.

To mix purple paint an artist uses 50 ml of blue paint and 125 ml of red paint.

The ratio of blue to red paint is 50 : 125.

Dividing both parts of the ratio by 50 gives: 1 : 2.5.

The relationship between the volume of blue and red paint can be described as:

'1 part blue to 2.5 parts red'.

Expressing a relationship as a ratio

If one number or quantity is a multiple of another, you can express the relationship as a ratio.

If the number of girls in the class is three times the number of boys then you can say:

- For every 3 girls there is 1 boy.
- The ratio of girls to boys is 3 : 1.

WORKIT!

A bag contains red and blue balls. $\frac{5}{8}$ of the balls are red.

Write down the ratio of red balls to blue balls.

The ratio of red to blue balls is 5 : 3 ← If $\frac{5}{8}$ of the balls are red and the rest are blue then $\frac{3}{8}$ must be blue.

NAIL IT!

The order in which the numbers are written matches the order in which the words are written.

- If you are told the ratio of boys to girls is 5 : 4 it means there are 5 boys for every 4 girls.

- If you are told the ratio of girls to boys is 5 : 4 it means there are 5 girls for every 4 boys.

DO IT!

What is the ratio of girls to boys in your class? Simplify the ratio if you can.

STRETCH IT!

The ratio of men to women in Bahrain is approximately 31 : 25.

What fraction of the population of the country is male?

DO IT!

Adrian says that if you know what fraction of people have a dog for a pet, you can easily write down the ratio of dog owners to people who don't own a dog.

Explain how he would do this.

Sharing in a given ratio

To share a quantity in a given ratio, split it into the number of 'parts' given by the ratio.

If Amy and Rudi share £90 in the ratio 4:5, think of the parts of the ratio as boxes to fill.

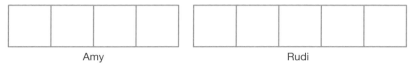

Amy	Rudi

There are 4 + 5 = 9 boxes to fill.

Each box must contain the same number.

£10	£10	£10	£10	£10	£10	£10	£10	£10

Amy	Rudi

90 ÷ 9 = £10

Amy receives 4 × £10 = £40.

Rudi receives 5 × £10 = £50.

NAILIT!

You can always check ratio calculations by adding together your answers to see if you reach the total you started with.

In the example on the left £40 + £50 = £90.

SNAPIT!   Sharing in a given ratio

To share in a given ratio:

1. Add the parts of the ratio together to find the total number of parts.
2. Divide the amount to be shared by the total number of parts.
3. Multiply this amount by each part of the ratio.

WORKIT!

To make a soft drink, cordial and water are mixed in the ratio 2:15.

A bottle contains 1 litre of cordial.

a How many litres of water are needed to mix the drink?

15 ÷ 2 = 7.5 litres ◀——

> The ratio 2:15 means that for every 2 litres of cordial you need 15 litres of water.
> If you only have 1 litre of cordial, divide by 2 to work out the volume of water required.

b How many litres of the drink will be made?

1 + 7.5 = 8.5 litres ◀——

> Add together the volume of the cordial and the water.

c How many litres of cordial will you need to mix 170 litres of the drink?

2 + 15 = 17

170 ÷ 17 = 10 ◀——

> Share 170 in the ratio 2:15 using the steps in the Snap it! box above.

Amount of cordial = 2 × 10 = 20 litres

Writing ratios as functions

If a ratio is in the form $1:a$, it can be written as a function.

For example, if the number of girls to boys is $1:2$ then for every 1 girl there are 2 boys.

You can write this as the equation $b = 2g$ where b = boys, g = girls.

If there are 4 girls, you could work out the number of boys by substituting $g = 4$ into the function:

$b = 2 \times 4 = 8$ boys

WORKIT!

A recipe asks for the number of eggs to the amount of sugar, in grams, in the ratio $1:120$.

Write down a function relating the number of eggs (e) to the amount of sugar (g).

$g = 120e$ ← The grams of sugar needed is 120 times the number of eggs.

CHECKIT!

1 Simplify each ratio.

  a $3:12$ **c** $8:10$

  b $5:15:20$

2 On a school trip there are 35 students and 5 teachers.

Write down the ratio of students to teachers. Simplify your answer.

3 A mix for concrete requires 250 g of cement to 375 g of sand. Copy and complete the statement: 'allow 1 part cement for … parts sand'.

4 A cinema sells 7 times more tickets in the evening than in the afternoon.

  a Write down the ratio of the number of tickets sold in the evening to the number of tickets sold in the afternoon.

  b In one day, 800 tickets were sold. How many were sold in the afternoon?

5 A survey suggests that $\frac{3}{5}$ of cats prefer Pawpaw cat food.

  a What is the ratio of cats that prefer Pawpaw cat food to those who don't?

  b The survey tested 200 cats. How many preferred Pawpaw cat food?

6 In a recipe the total volume of flour and sugar required is 1.5 kg.

If the ratio of flour to sugar is $2:3$, how much flour is needed?

7 The ratio of sides in a triangle is $3:4:5$.

If the shortest side has a length of 9 cm what are the lengths of the two longer sides?

8 The ratio of teachers to students in a school is $1:20$.

Write down a function relating the number of teachers (t) to students (s).

9 A recipe asks for sugar and flour in the ratio $2:5$.

If there is 60 g more flour than sugar, how much sugar is there in the recipe?

Scale diagrams and maps

Scale diagrams are used when you cannot draw things their actual size.

This is a map showing part of Hampshire.

The scale is 1 : 1 000 000.

This means 1 cm on the map represents 1 000 000 cm in real life.

On this map, the distance between Petersfield and Guildford is 3.5 cm.

© Petersfield Town Council (contains Ordnance Survey data
© Crown copyright and database right 2010)

To work out the distance in real life, multiply by 1 000 000.

Distance = 3.5 × 1 000 000 = 3 500 000 cm

Then convert this into km:

3 500 00 cm = 35 000 m ◄— Divide by 100 since 1 m = 100 cm.

35 000 m = 35 km ◄— Divide by 1000 since 1 km = 1000 m.

NAILIT!

To convert dimensions on a map to real life multiply them by the large number in the scale. Divide real life distances to convert them to map dimensions.

DOIT!

Using the map shown, work out the real-life distances between:

a Portsmouth and Chichester

b Southampton and Portsmouth

c Guildford and Lewes.

WORKIT!

The diagram shows the scale drawing of a ship and the harbour. How far and on what bearing is the ship from the harbour?

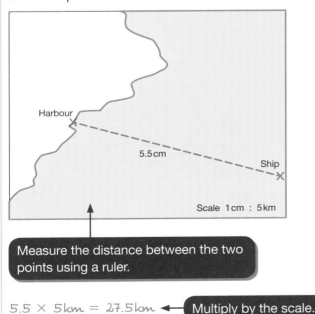

Measure the distance between the two points using a ruler.

5.5 × 5km = 27.5km ◄— Multiply by the scale.

Draw a North line at the harbour, then measure the bearing from North.

The bearing is 106°.

See page 128 for more on bearings.

101

If the scale on a map is 1: x, how would you work out the length in cm of a line representing 50 miles?

Estimating length using a scale

If you know the actual length of something shown in a scale drawing you can use that object to estimate the height of other objects.

In this diagram the beacon is approximately 3.5 times the height of the man.

Therefore, the height of the beacon = 3.5 × 1.8 = 6.3 metres.

1.8 m

CHECK**IT!**

1 The scale on a map says 1 : 50 000.

Which of the following is true? You may choose more than one answer.

A 1 cm represents 50 000 cm.

B 1 cm represents 0.5 km.

C 1 cm represents 50 000 m.

D 1 cm represents 5 km.

E 1 cm represents 5000 m.

F 1 cm represents 500 m.

2 A map has a scale 1 cm : 12 km.

 a The distance on the map between town *A* and town *B* is 3 cm. What is the distance in real life?

 b The distance between town *B* and town *C* is 15 km. How far apart are they on the map?

3 A scale model of a boat is made using a scale of 1 : 1000.

If the scale model is 12 cm long, how long is the original boat in metres?

4 The diagram shows an underground railway network.

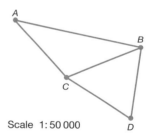

Scale 1 : 50 000

 a What is the distance in km between stations *A* and *C*?

 b What is the bearing from *B* to *C*?

Fractions, percentages and proportion

If there are 7 men and 3 women in a group, you could express the relationship between the number of men and women in several different ways.

- The **ratio** of men to women is 7:3.
- The **fraction** of men in the group is $\frac{7}{10}$.
- The **percentage** of men in the group is 70%.

Writing one quantity as a fraction of another

A fraction can be used to express a proportion of the whole, as in the examples below.

1 The shape has been split into 15 pieces; 6 of these are shaded.

So $\frac{6}{15}$ of the shape has been shaded.

2 7 can be written as a fraction of 10 by writing $\frac{7}{10}$.

3 9 cm can be written as a fraction of 1 metre by writing $\frac{9}{100}$. ◄────

> When writing one quantity as a fraction of another, make sure both quantities are in the same units!

4 36 hours can be written as a fraction of 1 day by writing $\frac{36}{24} = 1\frac{1}{2}$. ◄────

> Sometimes the fraction will be larger than 1.

Writing one quantity as a percentage of another

SNAPIT! **Writing one quantity as a percentage of another**

1 Write as a fraction.

2 Convert to a percentage.

DOIT!

Work out what proportion of your day you have spent doing maths work. Express the value as a fraction and as a percentage.

WORKIT!

Alin and Jackie share a prize in the ratio 2:3.

a What fraction does Jackie receive?

$\frac{3}{5}$ ◄──── Jackie receives 3 parts out of a total of 2 + 3 = 5 parts.

b What percentage does Alin receive?

$\frac{2}{5}$ ◄──── Alin receives 2 parts out of a total of 5 parts.

$= \frac{40}{100}$ ◄──── Multiply the numerator and denominator by 20 to get 100ths.

$= 40\%$

Proportion

When we talk about proportion we are thinking about what 'part' a number has in relation to a whole.

Proportions can be expressed as a fraction or a percentage.

For example, if there are 10 people in a room and 9 of them are male, the proportion of men is $\frac{9}{10}$ or 90%.

You can compare proportions by considering them as ratios – if the ratios are equal then the proportions are the same.

WORKIT!

A recipe for cake mix requires 120 g of sugar and 40 g of butter.

Granny says the recipe is wrong and you should use the rule '1 part butter to 3 parts sugar'.

> Write both proportions as ratios and compare the ratios. If the ratios are equal then the proportions are equal.

Show that the proportion of sugar to butter is the same in the recipe and using Granny's rule.

> Cancel the ratio down as much as possible.

(Recipe)   butter: sugar = 40 : 120 = 1 : 3

(Granny's rule) butter: sugar = 1 : 3

Since the ratios are equal, the proportions in both recipes are equal.

CHECKIT!

1 What fraction of 35 m is 20 cm?

2 During a day Alice spends:

2 hours riding

3 hours revising for exams

8 hours sleeping.

What fraction of the day does she have remaining for other activities?

3 In a box of 20 sweets, 15 are chocolates.

  a What fraction are chocolate?

  b What percentage of the sweets are **not** chocolates?

4 In a recipe, the ratio of butter to sugar to flour is 1 : 2 : 7.

What percentage of the mixture is butter?

5 The table below shows the number and gender of students at two schools.

	School A	School B
Boys	125	100
Girls	145	120

Is the proportion of girls to boys the same in both schools? Justify your answer.

6 A particular type of biscuit contains 22 g of fat in every 100 g. A 150 g packet of biscuits contains 8 biscuits.

If you eat one biscuit how many grams of fat do you consume?

Direct proportion

Two things are in direct proportion when they increase at the same rate.

For example, if you are paid £10 an hour, the number of hours you work is in direct proportion to how much you are paid.

If x and y are in direct proportion, an equation linking them is of the form $y = mx$, where m is a numeric value.

NAILIT!

Equations like $y = 2x$, $a = 3b$ and $m = 0.4n$ are all equations that show direct proportion.

Calculations involving direct proportion

To answer questions involving direct proportion, you must decide whether to multiply or divide. Think logically about whether you expect the values to increase or decrease.

There is often more than one way of solving problems involving proportion.

WORKIT!

For every 5 kg of plastic bottles a school recycles, they receive £32.

How much will they receive for 20 kg of plastic bottles?

Method 1

£32 ÷ 5 = £6.40 ◄— Work out how much the school receives for 1 kg of bottles.

20 × £6.40 = £128 ◄— Multiply, since you know they must receive **more** money for a larger volume.

Method 2

5 kg × 4 = 20 kg

£32 × 4 = £128 ◄— Since the weight has been multiplied by 4, the price must be multiplied too.

NAILIT!

It doesn't matter which method you use as long as you show your workings.

DOIT!

Decide which of the above methods you prefer for solving this problem.

Graphs showing direct proportion

A graph showing two things in direct proportion will be a straight line through the origin.

The equation of the line will always be in the form $y = mx$, where m is a constant.

For example, if you are paid £10 per hour for work then:

total earned = 10 × number of hours

(graph: y-axis labelled "Total earned (£s)" with values 0, 20, 40, 60, 80, 100, 120; x-axis labelled "Number of hours worked" with values 2, 4, 6, 8, 10; a straight line through the origin)

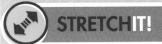

STRETCHIT!

A taxi driver charges a fixed fee of £2 plus £0.70 per mile. Explain why this is not an example of direct proportion.

Let y = total earned and x = number of hours worked.

The equation of the graph is $y = 10x$.

The gradient of the graph is 10, since for an increase of 1 in the number of hours worked the total earned will increase by £10.

You want to calculate how much the price increases for an increase in distance of 1 mile.

Choose two points on the graph and draw a right-angled triangle, marking on the increase in distance and price.

In this case, choosing the points (0, 0) and (10, 25) means you can easily see that for an increase of 10 miles an extra £25 is charged.

Divide £25 by 10 to find the price increase per mile.

WORKIT!

The graph shows the price of a taxi for different length journeys.

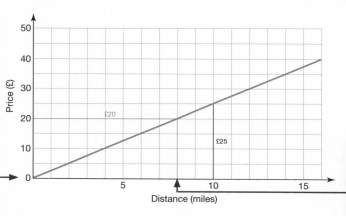

Draw a vertical line from 8 miles up to the graph and a horizontal line across to the y-axis, then read off the value.

a How much does it cost to travel 8 miles?

£20

b How much does each extra mile cost?

£2.50

c How much would a journey of 30 miles cost?

30 × £2.50 ◀── Each mile costs £2.50.

= £75

CHECKIT!

1 x and y are in direct proportion. Which of these could be an equation linking them?

A $y = 0.7x$

B $y = x + 0.7$ ── There may be more than one answer.

C $y = x - 0.7$

D $y = \frac{0.7}{x}$

E $y = 7x$

2 A recipe lists the ingredients for 20 meringues as 2 egg whites and 120 g sugar.

a How many meringues can you make with:

   i 3 eggs   **ii** 600 g of sugar?

b How many eggs would you need for 70 meringues?

3 A swimming pool fills at the rate of 4.5 litres per minute. How long will it take before the pool contains 675 litres? Give your answers in hours and minutes.

4 The graph below shows the relationship between the weight of flour and sugar needed for a recipe.

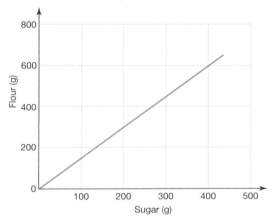

The gradient of the graph is 1.5. Which of the following statements are true?

A If you multiply the weight of sugar by 1.5 you will calculate the weight of flour.

B The weight of sugar is 1.5 times the weight of flour.

C For every 1.5 g of sugar you need 1 g of flour.

D For every 1 g of sugar you need 1.5 g of flour.

Inverse proportion

Two things are in inverse proportion if as one increases the other decreases at the same rate.

For example, as the speed at which you travel goes up, the time your journey will take goes down.

DOIT!

Decide if each of these situations describes things in direct proportion or inverse proportion:

a The number of people painting a fence and the time taken.

b The number of tickets purchased and the total price.

If x and y are in inverse proportion, an equation linking the two is of the form $y = \frac{m}{x}$, where m is a numeric value.

NAILIT!

Equations like $y = \frac{2}{x}$, $a = \frac{3}{b}$ and $m = \frac{0.4}{n}$ are all equations that show inverse proportion. Their graphs all look similar to this:

SNAPIT! Proportionality

If y is inversely proportional to x, then y is proportional to $\frac{1}{x}$.

Calculations involving inverse proportion

You should always think carefully about whether to multiply or divide. Are you expecting the value to increase or decrease?

WORKIT!

It takes 5 minutes to milk one cow using a milking machine. A farm purchases 4 milking machines. How long will it take to milk a herd of 100 cows?

Method 1

$5 \times 100 = 500$ minutes ◄──── Work out how long it would take 1 machine to milk 100 cows.

$500 \div 4 = 125$ minutes ◄──── If there are more machines it will take less time, so divide by 4.

$= 2$ hours 5 minutes

Method 2

$100 \div 4 = 25$ ◄──── They can milk 4 cows at a time, so divide by 4.

$25 \times 5 = 125$ minutes ◄──── Multiply by the number of minutes it will take.

$= 2$ hours 5 minutes

NAILIT!

Use your common sense to answer questions involving direct and inverse proportion.

For example, if you want to boil 100 eggs and each egg needs 3 minutes, you shouldn't boil the eggs for $3 \times 100 = 300$ minutes $= 5$ hours!

CHECKIT!

1 Two amounts, x and y are in inverse proportion. Which of these could be an equation that links them?

A $y = 0.7x$

B $y = x + 0.7$

C $y = x - 0.7$

D $y = \frac{0.7}{x}$

E $y = 7x$

2 If Hanisha travels at 60 miles per hour she can get to work in 15 minutes. How long will it take her if she travels at 40 miles per hour?

3 It takes two decorators five days to decorate the outside of a house. How long would it take six decorators?

4 The graph shows the age of chickens on a farm and the average number of eggs they lay every week.

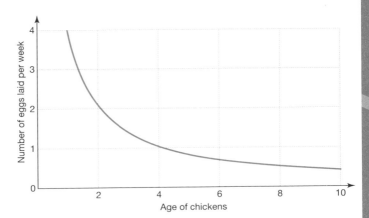

a How many eggs does a two-year-old chicken lay every week?

b Describe the relationship between the age of the chickens and the number of eggs they lay.

Working with percentages

Percentage change

When things increase or decrease you can calculate the percentage change.

If a price increases by 50% then the new price is 150% of the old price. Alternatively, you could say it is $1\frac{1}{2}$ times the original value or 1.5 times the original value.

> Percentage change can also be expressed as a fraction or a decimal.

WORKIT!

The number of students taking Mandarin at A level in 2017 is 75% of the number who took it in 2016.

Which of the following statements are true? Explain your answers.

a The number of people taking Mandarin has decreased by 25%.

True – 75% is 25% less than 100%.

b The number of people taking Mandarin in 2017 is 0.25 times the number who took it last year.

False – the number of people taking Mandarin in 2017 is 0.75 times the number who took it in 2016 since 75% = 0.75.

c The number of people taking Mandarin in 2017 is $\frac{3}{4}$ of the number of people who took it in 2016.

True – The number of people taking Mandarin in 2017 is 75% of the number who took it in 2016. 75% = $\frac{3}{4}$.

WORKIT!

The price of juice has risen from 89p per litre to £1.25 per litre. Work out the percentage change. Give your answer to the nearest percentage.

Increase in price = 125 – 89 = 36p

$\frac{36}{89} \times 100 = 40\%$ increase

SNAPIT! Calculating percentage change

1 Find the change in value by subtracting.

2 Divide the change by the original value.

3 Multiply by 100.

Comparing two quantities using percentages

We can compare two quantities by considering what percentage one is of the other.

For example, if two different companies sell a DVD at different prices:

Company A = £15   Company B = £12

We could say that Company A charges: $\frac{15}{12} \times 100 = 125\%$ of the price Company B charges.

We could also say that Company B charges: $\frac{12}{15} \times 100 = 80\%$ of the price Company A charges.

Percentage profit or loss

Profit or loss can be calculated in exactly the same way as percentage change.

Calculating original values

If you are given the value after an increase or decrease, you may need to find the original value.

For example, after a 12% increase in price a DVD player costs £56.

112% of the original price = £56

1% of the price = £56 ÷ 112 = £0.50

The original price is 100%:

100 × £0.50 = £50

WORKIT!

In a sale, all prices are reduced by 40%. If the new price of a video game is £18, what was the original price?

60% of the original price = £18

1% of the original price = £18 ÷ 60 = £0.30

100% of the original price = £30.00

SNAP IT!   Finding the original price

1. Write down the percentage of the price you are given.

2. Divide the price by the percentage to work out 1%.

3. Multiply by 100 to give the original price.

STRETCH**IT!**

A credit card charges 28% APR (annual percentage rate). This means if you do not pay off the money owed, the bank will charge an additional 28% per year.

If you spend £100 how much will you have to pay the credit card company back?

Calculating interest

When you save money in a bank you will often be paid **interest**, at a stated percentage rate, if you leave the money in the account. This is usually calculated annually (at the end of the year).

To work out how much money will be in the account, calculate:

total increase + original savings

When you only receive interest on the initial investment, this is called **simple interest**.

For example, if you put £100 in a bank with 5% simple interest, then each year you will receive £5.

WORKIT!

A bank offers 2.5% interest for a savings bond. Paul invests £3500.
If he does not withdraw any money, how much will he have at the end of the year?

$$2.5\% \text{ of } £3500 = 0.025 \times 3500$$
$$= £87.50$$

$$£3500 + £87.50 = £3587.50$$

Growth and decay

If something increases year on year, it experiences **growth**.
If something decreases year on year, it experiences **decay**.
For example, the population of the UK increases by approximately 5% each year.

If the population in 2016 is approximately 64 million, you can predict the population over the next few years in the following way:

2017   105% of 64 000 000 = 1.05 × 64 000 000 = 67 200 000
2018   105% of 67 200 000 = 1.05 × 67 200 000 = 70 560 000
2019   105% of 70 560 000 = 1.05 × 70 560 000 = 74 088 000

... and so on.

To work out what the population would be after 10 years, multiply by 1.05 ten times.

A quicker way to do this is to calculate $1.05^{10} \times 64\,000\,000 = 104\,249\,256$.

WORKIT!

The number of bacteria in a petri dish decreases by 10% each hour.

There are 32 000 bacteria in the dish. How many will there be after:

a 1 hour

$$0.9 \times 32\,000 = 28\,800$$

b 5 hours

$$0.9^5 \times 32\,000 = 18\,896$$

You could multiply by 0.9 five times, but it is quicker to multiply by 0.9^5.

Round your answer to a sensible degree of accuracy, in this case the nearest whole number.

If the number of bacteria decreases by 10% there will be 90% left, so find 90% of 32 000.

Compound interest

When you receive interest on your interest in a bank account, this is called **compound interest**.

You calculate compound interest in exactly the same way as any other growth.

For example, if you invest £300 in a bank account with 3% compound interest you could work out how much you would have after 10 years as shown on the right.

The number of years you have invested the £300 for.

$$£300 \times 1.03^{10} = £403.17$$

Original investment.

Multiply by 1.03 since you wish to find 103% of the original investment.

STRETCHIT!

You invest £100 in a bank account. What percentage interest would you need to have if you wish to have more than £110 after 5 years?

SNAPIT! You need to know some financial terms

Profit	How much money is made
Loss	How much money is lost
Cost price	The price raw materials cost to purchase
Selling price	The price a product is sold for
Debit	Money that you pay or owe – if your bank balance is £100 in debit it means you owe the bank £100
Credit	Money that you earn, receive or possess – if your bank account has £100 credit it means you have £100 to spend
Income tax	The amount you pay the government when you earn money, worked out according to the amount you earn
VAT	Value Added Tax – the tax the government adds to the price of 'luxury' items; it is currently 20% in the UK
Interest rate	How much money you will be paid if you leave your investments in a bank account or similar; it is usually a percentage of the money you invest, paid annually or over some other time frame

CHECKIT!

If you borrow money you are the one who has to pay interest – to the bank. The percentage you pay is usually higher than the rate the bank pays to its investors. This is how banks make a profit.

1 Work out
   a 103% of £50   c 19.5% of 64
   b 248% of 400

2 A music teacher puts his prices up from £40 per hour to £45 per hour. Work out the percentage increase.

3 The average temperature in a Mediterranean country one year was 24°C, 15% greater than in the previous year. What was the average temperature in the previous year?

4 The number of hits on a website increases at a rate of 20% per hour. Initially there are 15 000 hits. How many will there have been after 3 hours?

5 In a sale the price of a TV was reduced by 20%. If it was reduced by £30, what was the original price?

6 In Year 1 a company has 225 employees. In Year 2 there are only 200.

Copy and complete the statement:

The number of employees in Year 2 is …% of the number in Year 1.

Compound units

A compound measure is made up of two or more units.

Examples of compound units include speed, density, pressure and rates of pay.

Speed can be measured in, for example, km/h, m/s or mph (miles per hour).

Density can be measured in, for example, g/cm^3 or kg/m^2.

WORKIT!

A call centre charges 13p per minute for calls. How much will a half hour call cost?

13p × 30 mins = 390p

= £3.90

> Convert $\frac{1}{2}$ hour into minutes.

WORKIT!

How far will a tram travelling at 75 km/h travel in 20 minutes?

Distance = speed × time

Speed = 75km/h

Time = 20 minutes = $\frac{1}{3}$ hour

Distance = 75 × $\frac{1}{3}$

= 25km

> Convert the time into hours so the units are the same.

STRETCHIT!

If you travel 100 miles in x hours, write an expression to describe your speed in miles per hour.

DOIT!

Write down the three formulae which connect pressure, force and area.

Rates of pay and prices

When working with compound units involving rates of pay and prices, make sure you are working in the same units.

Speed

The relationship between speed, distance and time is illustrated in the triangle.

Cover the unit you wish to calculate:

Speed = distance ÷ time

Distance = speed × time

Time = distance ÷ speed

Density

The relationship between density, mass and volume is illustrated in this triangle.

Density = mass ÷ volume

Mass = density × volume

Volume = mass ÷ density

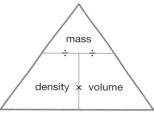

WORKIT!

Work out the density of an object that weighs 50 kg and has a volume of 32 m³. Give your answer to 2 decimal places.

Density = mass ÷ volume

= 50kg ÷ 32 m³

= 1.56kg/m³

> The unit is kg/m³ since the original units were kg and m³.

Pressure

The relationship between pressure, force and area is illustrated in this triangle.

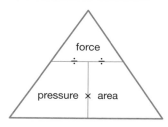

WORKIT!

The pressure on a plate is 35 N/m². If the force is 70 N, what is the area of the metal plate?

Area = force ÷ pressure

   = 70 N ÷ 35 N/m²

   = 2 m²

> Since the force is in Newtons (N) and the pressure is Newtons per m² (N/m²) you do not need to convert units.

DOIT!

Practise drawing out the triangles illustrating the relationships between speed/distance/time, density/mass/volume and pressure/force/area units.

Converting compound units

When converting compound units, convert one unit at a time.

For example, to convert 18 km per hour into metres per second:

18 km per hour = 18 000 m per hour   ← There are 1000 m in 1 km.

18 000 m per hour = 18 000 m per 60 minutes   ← There are 60 minutes in 1 hour.

   = 300 m per minute   ← Divide by 60 to work out how many m per minute.

300 m per minute = 300 m per 60 seconds   ← There are 60 seconds in 1 minute.

   = 5 m per second   ← Divide by 60 to work out how many m per second.

WORKIT!

A sewing machine sews at 1 inch per second. Work out the speed in metres per minute.

1 inch = 2.5 cm   ← First convert into cm.

1 inch per second = 2.5 cm per second

2.5 × 60 = 150 cm per minute   ← Work out how many cm it will sew in 1 minute.

   = 1.5 m per minute   ← Convert into metres.

NAILIT!

A shortcut to convert from km per hour to m per second is to divide by 3.6.

✓ CHECKIT!

1 Electricity costs 18p per unit. If the bill at the end of the month is £29.50, how many units of electricity have been used? Give your answer to the nearest whole number.

2 A train travels at 120 miles per hour. How long will it take to travel 80 miles? Give your answer in minutes.

3 The volume of a cork is 3 cm³. It weighs 0.72 g. What is the density?

4 A force of 12 N is applied to an area of 2 m². Calculate the pressure.

5 Convert 3 m/s into km/h.

6 A petrol pump flows at 0.6 litres per second. Convert the rate of flow into gallons per hour.

1 gallon = 4.55 litres

7 Usain Bolt ran 100 m in 9.58 seconds. The top speed of a cheetah is 120 km/h.

Who is faster? Show all your calculations.

🚫 You may **not** use a calculator for these questions.

1 Work out how many

   a m in 3.2 km

   b seconds in 9 minutes

   c ml in 0.4 litres.

2 Convert 4600 m into km.

3 How many minutes are there in 2.5 hours?

4 Calculate the area of this shape.

1.1 m

32 cm

5 A square has area 3 m². What is its area in cm²?

6 What fraction of 12 is 5?

7 A bag contains 26 blue balls and 18 red balls. Write the ratio of blue balls to red balls. Give the ratio in its simplest form.

8 Boiling water in a kettle cools at approximately 3° C per minute. To make a cup of tea, water should be a maximum of 85° C. How long should you wait after a kettle has boiled before making a cup of tea?

9 A metal object has mass 345 kg and volume 0.15 m³. Calculate the density of the object.

10 One day Ian runs 8 km. The next day he runs 10 km. Work out the percentage increase in the distance he runs.

11 In a restaurant, 13 of the 25 customers order the set menu. Work out the proportion of customers who do not order the set menu.

12 A DJ runs a radio programme between 9.00 am and 10.30 am. Here is the schedule.

9.00 am	News and Weather
9.15 am	80s Music
10.00 am	News
10.05 am	Weather
10.08 am	Quiztime
10.20 am	Showbiz Roundup

What proportion of the schedule is taken up with news and weather?

13 A snail crawls at $\frac{4}{5}$ of a metre per hour. How long will it take to travel the length of a garden measuring 20 m?

14 Josie and Charlie share 50 marbles in the ratio 1 : 4.

   a How many more marbles does Charlie have than Josie?

   b Write a function relating the number of marbles Josie has (J) to the number Charlie has (C).

🖩 You **may** use a calculator for these questions.

15 What percentage of this circle is shaded? Give your answer to 2 significant figures.

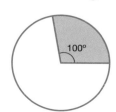

100°

16 In a shop the normal price of a sofa is £1200. In a pre-Christmas sale all prices are reduced by 20%.

After Christmas a further 10% is taken off sale prices.

What is the price of the sofa now?

17 A map uses a scale of 1 : 50 000. A particular road is 3 km long. Work out the length of the road on the map. Give your answer in cm.

18 Henry invests £1500 in a savings account. He is paid 2% per year in compound interest.

How much will Henry have in the account at the end of 3 years?

19 One in four people who are contacted by a charity donate to the charity. If the charity contacts 32 000 people in a year, how many will donate?

20 The train track from London to Edinburgh is 393 miles long. If a train travels at a speed of 125 mph, how long will the journey take? Give your answer in hours and minutes.

21 A café charges the following prices:

Coffee   £2.50

Tea   £1.90

Sandwiches   £5.30

All prices are increased by 5%.

Mr Angle always buys 1 coffee, 1 tea and 2 sandwiches. What is the new price of his order?

22 The number of people using the facilities in a sports centre at 10.00 am one morning is recorded in a table as follows.

Facility	Number of customers
Swimming pool	37
Gym	15
Squash courts	4
Café	19

What percentage of the customers are using the gym?

23 A savings account offers 4.5% simple interest.

If Allan puts £3000 into an account, how much will he have in the account at the end of 5 years?

24 There are 30 children in a class. The ratio of boys to girls is 2:1. Two boys leave and another three girls join the class. What is the ratio of boys to girls now?

25 A concert is attended by 3450 people.

The ratio of men to women is 7:6.

The ratio of women to children is 15:2.

How many men are at the concert?

26 A mobile phone company uses this graph to calculate the cost of mobile phone bills.

Erin says 'the cost of the bill is in direct proportion to the number of minutes of calls you make'.

Is she correct? Explain your answer.

27 The time to cook a turkey is given as 25 minutes per 450 g, plus an extra 25 minutes.

Is the relationship between the weight of the turkey and the time taken to cook it:

  i direct proportion

 ii inverse proportion

iii neither?

Explain your answer.

28 A crocodile can run at a speed of 17 km per hour over short distances. How long will it take to run 50 m? Give your answer to the nearest second.

29 The table shows the people using a hotel on one day.

	Adults	Children
Female	35	12
Male	60	37

Alex says 'the proportion of adults and children who are male is the same, since $37 - 12 = 25$ and $60 - 35 = 25$'. Is she correct? Explain your answer.

30 The ratio of butter to sugar in a recipe is 1 part butter to 2.5 parts sugar.

Mandeep uses 90g of butter.

How much sugar should she use?

Geometry and measures

Measuring and drawing angles

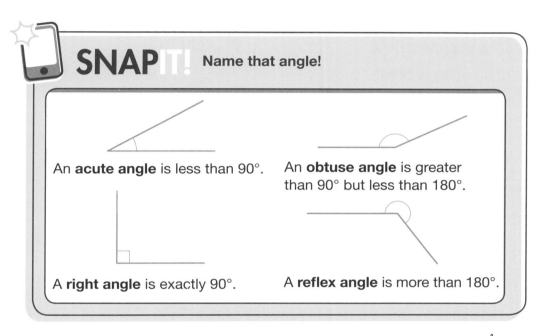

An **acute angle** is less than 90°.

An **obtuse angle** is greater than 90° but less than 180°.

A **right angle** is exactly 90°.

A **reflex angle** is more than 180°.

You can use letters to name angles. This is angle *ABC*.

You use a **protractor** to measure and draw angles.

The angle is always at the middle letter.

Measuring angles

To measure an angle:

Outside scale

Inside scale

1 Place the cross at the centre of the protractor over the point of the angle.

2 Line up the base line of the protractor with one line of the angle.

3 Use the outside scale to measure clockwise turns. Use the inside scale to measure anticlockwise turns.

NAILIT!

Always estimate the size of the angle before you measure it.

WORKIT!

Measure reflex angle *DEF*.

Smaller angle = 115°

Angle DEF = 360° − 115° = 245°

> Measure the smaller angle using the inside scale (start from 0°).
> Subtract the smaller angle from 360°. (See angle facts on page 121.)

Drawing angles

To draw an angle of 36°:

- Draw a horizontal line with a ruler.

- Place the cross at the centre of the protractor on one end of the line. The base line on the protractor must rest on the line.

- Using the inside scale (starts from 0°), make a mark at 36°.

- Join the mark to the end of the line with a ruler.

- Label the angle.

DOIT!

Draw and label angle
ABC = 140°.
Mark on your diagram
angle *ABC* = 220°.

CHECKIT!

1 a Measure the size of the angle marked *x*.

   b What type of angle is this?

2 Draw and label these angles.

   a angle *XYZ* = 118°

   b angle *ABC* = 290°

   c angle *LMN* = 42°

3 Here is triangle *PQR*. This diagram is accurately drawn.

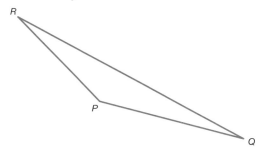

   a Mark with a letter *O* an obtuse angle.

   b Measure the size of the angle *QRP*.

Using the properties of angles

Angle facts

You need to remember these angle facts.

SNAP IT!   Angle facts

The angles on a straight line add up to 180°.	The angles around a point add up to 360°.	Vertically opposite angles are equal.

$a + b = 180°$

$c + d + e = 360°$

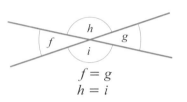

$f = g$
$h = i$

WORKIT!

a

Write down the value of a.

$a = 45°$

Vertically opposite angles are equal.

b

This diagram is wrong. Explain why.

$115° + 115° + 150° = 380°$
Angles around a point should add up to 360°.

DO IT!

There are three angles x, y, z at a point. One is acute, one is obtuse and one is reflex. Write down one possible set of the three angles x, y, z.

SNAP IT!   Special triangles

An **isosceles triangle** has two equal sides and two equal angles.	An **equilateral triangle** has three equal sides and all angles equal to 60°.	A **right-angled triangle** has one angle equal to 90°.

In a **scalene triangle** none of the sides or the angles are the same.

Angles in triangles

The **interior angles** of a triangle add up to 180°.

$a + b + c = 180°$

A **vertex** (plural: **vertices**) is a point where two edges meet.

An **exterior angle** of a triangle is equal to the sum of the interior angles at the other two **vertices**.

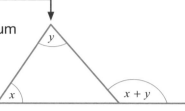

You can use letters to name lines and the sides of shapes. Here are line *BC* and triangle *DEF*.

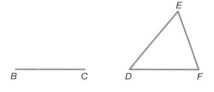

NAILIT!

The **base angles** of an isosceles triangle are the equal angles.

DOIT!

One angle in an isosceles triangle is 70°. Show that there are two possible solutions for the other two angles.

One angle in an isosceles triangle is 72°. Work out the size of the smallest possible angle in the triangle.

WORKIT!

$BA = BC$

a Work out the size of the angle marked *b*.

Angle $b = 180° - 55° - 55° = 70°$

b Give a reason for your answer.

Triangle ABC is isosceles.

Angle BAC = angle BCA.

Write angle *BCA* on the diagram.

SNAPIT! **Opposite angles of a parallelogram**

$e = g$

$f = h$

See page 131 for more on the properties of quadrilaterals.

Angles in quadrilaterals

The interior angles of a quadrilateral add up to 360°.

$a + b + c + d = 360°$

WORKIT!

ABCD is a quadrilateral.

CDE is a straight line.

Work out the size of angle *x*.

Give reasons for your answer.

Angles in a quadrilateral add up to 360°.

Angle $ADC = 360° - 108° - 122° - 65° = 65°$

Angles on a straight line add up to 180°.

Angle $x = 180° - 65° = 115°$

Parallel and perpendicular lines

Perpendicular lines meet at 90°. In this diagram, line *PQ* is perpendicular to line *RS*.

Parallel lines are lines that always remain the same distance apart. On diagrams they are marked with arrows.

You need to remember these angle facts about parallel lines.

SNAP IT!   Angle facts about parallel lines

Corresponding angles are equal.

Alternate angles are equal.

Co-interior angles (or **allied angles**) add up to 180°.

$a + b = 180°$

WORKIT!

BC is a straight line.

Is the line *AB* parallel to the line *DC*?

Give a reason for your answer.

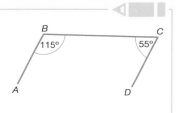

In parallel lines co-interior angles add up to 180°.

$115° + 55° = 170°$

The line *AB* is not parallel to the line *DC*.

DOIT!

Here is parallelogram *ABCD*.

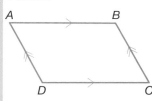

a Draw a line through *B* perpendicular to *DC*.

b Draw a line through the midpoint of *AB* parallel to *CB*.

Solving angle problems

When solving more involved angle problems you will need to use two or **more** angle properties. Make sure you give a **clear** reason for each step of your working.

DOIT!

Create a poster to show each of these angle properties you will use in angle problems.

a Angles on a straight line add up to 180°.

b Angles around a point add up to 360°.

c Vertically opposite angles are equal.

d Angles in a triangle add up to 180°.

e Base angles of an isosceles triangle are equal.

f Angles in a quadrilateral add up to 360°.

g Opposite angles of a parallelogram are equal.

h Corresponding angles are equal.

i Alternate angles are equal.

WORKIT!

PQR is a straight line.

PT = QT

You can tell from the diagram that PTS is **not** a straight line since at T: 80° + 75° = 155° (and the angles on a straight line add up to 180°).

a Work out the size of angle a.

a = angle TPQ   (Base angles of an isosceles triangle are equal.)

a = (180° − 80°) ÷ 2 = 50°   (Angles in a triangle add up to 180°.)

b Work out the size of angle b.

Angle TQR = 180° − 50° = 130°   (Angles on a straight line add up to 180°.)

b = 360° − 130° − 75° − 112° = 43°   (Angles in a quadrilateral add up to 360°.)

You also need to be able to 'show that' something is true.

WORKIT!

ABCD is a parallelogram.

Show that triangle ECB is an isosceles triangle.

You are told that ECB is an isosceles triangle, so the base angles will be equal.
Work out all the angles of triangle ECB to show that two of the angles are equal.

Angle ECB = 50°   (Opposite angles of a parallelogram are equal.)

Angle CEB = 180° − 115° = 65°   (Angles on a straight line add up to 180°.)

Angle CBE = 180° − 50° − 65° = 65°   (Angles in a triangle add up to 180°.)

Base angles of an isosceles triangle are equal.

Angle CEB = angle CBE so triangle ECB is an isosceles triangle.

NAILIT!

The question could have been set as: 'ABCD is a parallelogram. What type of triangle is triangle ECB?' Your working and conclusion would be exactly the same.

NAILIT!

For some angle problems, you may be given the angles in terms of letters and numbers, for example x + 24°. You will need to set up and solve equations to find the missing angles. If you're unsure how to do this, have a look at page 55.

CHECKIT!

1 Work out the size of angle a.

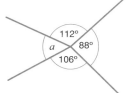

2 *XYZ* is a straight line.

$WX = WY$

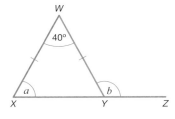

a Work out the size of angle a.
Give a reason for your answer.

b Work out the size of angle b.

3 Work out the value of x.

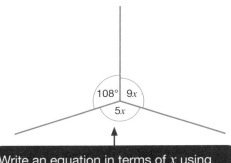

Write an equation in terms of x using the fact that angles around a point add up to 360°.

4 *RST* is a straight line.

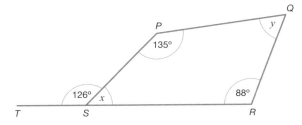

a Work out the size of angle x.
Give a reason for your answer.

b Work out the size of angle y.

5 *AB* is parallel to *CD*. *EH* is a straight line.

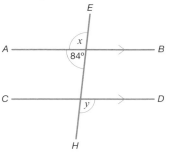

a Work out the size of angle x.

b Work out the size of angle y.
Give a reason for your answer.

6 *PQRS* is a parallelogram.

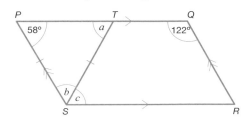

a Work out the size of angle a.

b Work out the size of angle b.

c Work out the size of angle c.
Give a reason for your answer.

7 *AB* is parallel to *DC*. *DA* is parallel to *CB*.
AE = DE. *EDC* is a straight line.

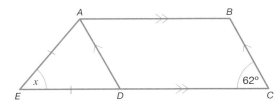

Work out the size of angle x.
Give reasons for your answer.

8

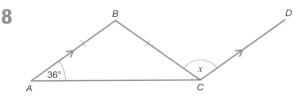

AB is parallel to *CD*. *AB = CB*.

Work out the size of angle x. Give reasons for each stage of your working.

Using the properties of polygons

SNAP IT! Name that polygon!

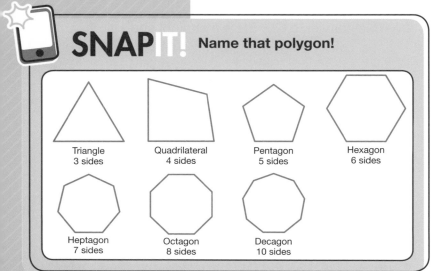

Triangle
3 sides

Quadrilateral
4 sides

Pentagon
5 sides

Hexagon
6 sides

Heptagon
7 sides

Octagon
8 sides

Decagon
10 sides

A **polygon** is a 2D shape with three or more straight sides.

In a **regular polygon** all the sides are equal and all the angles are equal; otherwise it is an **irregular polygon**.

You need to be able to recognise and name the polygons with 3 to 8 sides and the polygon with 10 sides.

A regular triangle would be an equilateral triangle (see page 121), and a regular quadrilateral would be a square (see page 131).

STRETCH IT!

1 A hexagon can be divided into four triangles when all the diagonals are drawn from one vertex (point).

Complete these statements:

The angle sum of a triangle is _____°

Sum of interior angles of a hexagon = 4 × _____° = _____°

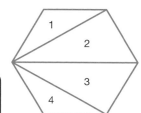

2 Think about dividing other polygons in the same way. Complete this table.

Polygon	Number of sides (*n*)	Number of triangles formed	Sum of interior angles
Triangle	3	1	180°
Quadrilateral	4	2	
Pentagon	5		
Hexagon			720°
Heptagon		5	
Octagon			1080°
Decagon			

3 How many triangles are formed for an *n*-sided polygon?

4 Write down the sum of the interior angles of an *n*-sided polygon.

The sum of the **interior angles** of a polygon with n sides is given by: $180° × (n − 2)$

For the polygon shown, $n = 5$.

So, sum of interior angles $= 180° × (5 − 2)$

$= 180° × 3 = 540°$

Interior angle   Exterior angle

You can use the fact that the angles on a straight line add up to 180° to work out an **exterior angle** if you know the interior angle, **or** to find the interior angle if you know the exterior angle.

At a vertex, interior angle + exterior angle = 180°

The sum of the exterior angles of any polygon is 360°.

WORKIT!

Here is a regular pentagon.

a Work out the size of angle x.

$x = 360° \div 5 = 72°$

b Work out the size of angle y.

$y = 180° - 72° = 108°$

x is an exterior angle. The exterior angles of a polygon add up to 360°. If a regular polygon has n sides, each exterior angle is $360° \div n$.

y is an interior angle. Interior angle + exterior angle = 180°

If you know the exterior angle of a regular polygon, you can work out how many sides the polygon has.

WORKIT!

The interior angle of a regular polygon is 162°.

How many sides does the polygon have?

Exterior angle $= 180° - 162° = 18°$ ← First work out the exterior angle.

Number of sides $= 360° \div 18° = 20$ ← The polygon is regular so all the exterior angles are 18°, with sum 360°.

 ## STRETCHIT!

Regular hexagons can fit together to form a pattern without any gaps or overlaps.

A square and an octagon can also fit together to form a pattern.

By considering the sum of angles at points A and B, give a reason why these polygons fit together as shown.

Can you fit together regular pentagons to form a pattern without any gaps or overlaps? Give a reason for your answer.

CHECKIT!

1 Work out the size of the exterior angle of a regular decagon.

2 The exterior angle of a regular polygon is 15°.

  a How many sides does the polygon have?

  b Work out the sum of the interior angles of the polygon.

3 The diagram shows a regular pentagon *ABCDE*.

The line *EB* is drawn.

Work out the size of angle *CBE*.

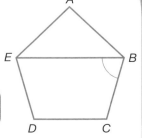

What can you say about triangle *ABE*?

Using bearings

A **bearing** is an angle measured clockwise from North, used to describe a direction.

A bearing is always written using three figures.

Bearing of *B* from *A* is 065°

> Add a zero at the start so that the bearing has three figures.

Bearing of *B* from *A* is 234°

> Bearings are measured clockwise from North. You need to complete a $\frac{1}{2}$ turn (180°) then an extra 54°:
>
> $180° + 54° = 234°$

Bearing of *B* from *A* is 125°

> Subtract 235° from a complete turn of 360°:
>
> $360° - 235° = 125°$

You should know how to 'measure the bearing of...'. The term 'measure' is telling you to use a protractor. If you're unsure about using a protractor, have a look at page 119.

You can work out a reverse bearing by adding or subtracting 180°.

NAIL IT!

Bearing of *B* from *A*

'From *A*' means that you are 'at *A* facing North'.

You need to work out what angle you need to turn through to face *B*.

DO IT!

Compass points

Draw the eight points of a compass (N, NE, E, SE, S, SW, W and NW). Add the three-figure bearing of each point (for example, N 000°, E 090°).

WORKIT!

The bearing of *A* from *B* is 072°.

Work out the bearing of *B* from *A*.

Bearing of *B* from *A* is 180° + 72° = 252°

> The marked angles are alternate angles. See page 123 for more on alternate and corresponding angles.

To solve a bearings problem you may need to give a bearing between the points on a map or scale drawing (see page 101). You also need to be able to draw an accurate diagram.

WORKIT!

The diagram shows the positions of two buildings, P and Q, on a map.

A building R is on a bearing of 055° from building P.

Building R is also on a bearing of 280° from building Q.

Draw an accurate diagram to show the position of building R.

Label it R.

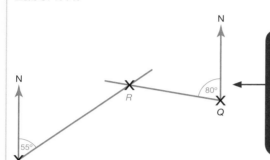

> Draw a line on a bearing of 055° from P.
> Draw another line on a bearing of 280° from Q.
> R must lie on both these lines.
> Mark where the lines cross with a cross (x) and label it R.

✓ CHECKIT!

1

a Measure and write down the bearing of B from A.

b On the diagram, draw a line on a bearing of 115° from A.

2 The bearing of Q from P is 164°.

Work out the bearing of P from Q.

> Draw a sketch of P and Q.

3 The diagram shows the positions of two boats, A and B.

Nathan says, 'The bearing of boat B from boat A is 046°'.

Kirsty says, 'The bearing of boat B from boat A is 314°'.

Who is correct? Give a reason for your answer.

Properties of 2D shapes

Lines of symmetry

A **line of symmetry** (or **mirror line**) is a line through a 2D shape such that each side is a mirror image of the other.

You should know how to complete the drawing of a shape so that it has line symmetry.

You also need to be able to shade more squares to complete a drawing so that it has line symmetry.

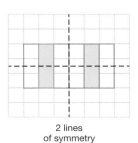

No lines
of symmetry

2 lines
of symmetry

> Each vertex (corner) of the shape has a mirror image the same distance away from the line of symmetry but on the other side of it.

1 line of symmetry

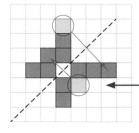

Rotational symmetry

A shape has **rotational symmetry** if it can be turned and fitted onto itself within one full turn.

The **order of rotational symmetry** tells you how many times the shape fits onto itself.

If a shape fits onto itself **only once** in a full turn it does not have rotational symmetry.

A shape that has rotational symmetry of order 2 looks the same after it has been rotated through a half turn.

No rotational
symmetry

Rotational symmetry
of order 2

> Each shaded square must have a mirror image if the pattern has the given line of symmetry.
> By comparing the two sides of the pattern, you can identify the squares that need to be shaded.

NAILIT!

You can use tracing paper to help you check for lines of symmetry and rotational symmetry. Trace the shape.

- Fold it in half along the line of symmetry. Check that both halves match up exactly.
- Rotate the tracing paper to see how many times the shape fits onto itself in one full turn.

DOIT!

Rotating polygons

1 Find the order of rotational symmetry of a
   a regular triangle (equilateral triangle)
   b regular quadrilateral (square)
   c regular pentagon
   d regular hexagon.
2 What is the link between the order of rotational symmetry and the number of sides of a regular polygon?

Quadrilaterals

A **quadrilateral** is any four-sided shape.

You need to know (and remember!) the properties of these special quadrilaterals.

SNAP IT! Quadrilaterals

Square
- All sides equal in length
- All angles are 90°
- Diagonals are equal and bisect each other at 90°
- 4 lines of symmetry
- Rotational symmetry of order 4

Rectangle
- Opposite sides equal in length
- All angles are 90°
- Diagonals are equal and bisect each other
- 2 lines of symmetry
- Rotational symmetry of order 2

Parallelogram
- Opposite sides are parallel and equal in length
- Diagonals bisect each other
- No lines of symmetry
- Rotational symmetry of order 2

Trapezium
- One pair of opposite sides are parallel

Kite
- Two pairs of **adjacent** sides are equal in length
- One pair of opposite angles are equal
- Diagonals cross at 90°
- 1 line of symmetry

Rhombus
- All sides equal in length
- Opposite sides are parallel
- Diagonals bisect each other at 90°
- 2 lines of symmetry
- Rotational symmetry of order 2

Make sure that you can:
- write down the name of a quadrilateral given in diagram form
- draw any of the quadrilaterals on a squared grid
- name the quadrilaterals that have a specific property
- complete a quadrilateral, given information about the shape, on coordinate axes (see page 73).

NAIL IT!

A line that joins the opposite corners of a quadrilateral is called a **diagonal**.

Bisect means 'to divide into two equal parts'.

Adjacent means 'next to'.

DO IT!

Name that quadrilateral!

Using presentation software, create a slideshow for each of the quadrilaterals. Put a different property on each slide. The final slide should give the name of the quadrilateral, together with the diagram.

How many slides do you need to identify each quadrilateral?

CHECK IT!

1 a Complete this shape so that it has line symmetry.

b Write down the name of the complete shape.

2 Here is a regular octagon.

a Draw a line of symmetry on the octagon.

b What is the order of rotational symmetry of the octagon?

3 Complete the following sentences.

a A rectangle has rotational symmetry of order _____.

b A _____ has all sides equal and rotational symmetry of order 2.

c A kite has _____ line of symmetry and _____ rotational symmetry.

d The diagonals of a _____ and a _____ bisect each other at 90°.

Congruent shapes

Shapes are **congruent** if they have exactly the same shape and size.

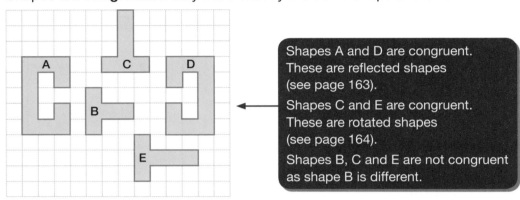

Shapes A and D are congruent. These are reflected shapes (see page 163).

Shapes C and E are congruent. These are rotated shapes (see page 164).

Shapes B, C and E are not congruent as shape B is different.

If two triangles are congruent their angles are the same and corresponding sides are the same length.

The diagram shows a pair of congruent triangles X and Y.

If you rotated triangle Y it would fit exactly onto triangle X.

Since triangles X and Y are congruent, you can write down that the length of side a is 8 cm and the size of angle b is 60°.

Two triangles are congruent when one of four conditions is true.

If any one of these conditions is true, the other will automatically be true too.

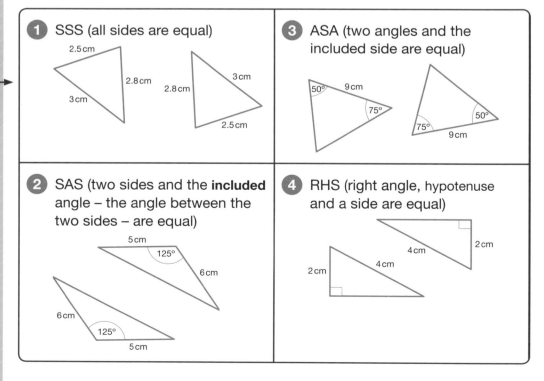

1. SSS (all sides are equal)
   2.5 cm, 2.8 cm, 3 cm, 2.8 cm, 2.5 cm, 3 cm

2. SAS (two sides and the **included** angle – the angle between the two sides – are equal)
   5 cm, 125°, 6 cm, 6 cm, 125°, 5 cm

3. ASA (two angles and the included side are equal)
   50°, 9 cm, 75°, 75°, 9 cm, 50°

4. RHS (right angle, hypotenuse and a side are equal)
   4 cm, 2 cm, 2 cm, 4 cm

NAILIT!

The **hypotenuse** is the longest side and is opposite the right angle in a right-angled triangle.

WORKIT!

Give a reason why this pair of triangles is congruent.

Label the sides (S) and the angles (A or R, for right angle) that you know.

Check to see if you have SSS, SAS, ASA or RHS.

ASA

SNAPIT! Congruent triangles

SSS = side, side, side
ASA = angle, side, angle
SAS = side, angle, side
RHS = right angle, hypotenuse, side

CHECKIT!

1 On a grid, draw a shape that is congruent to shape A.

2 Here is a pair of congruent triangles.

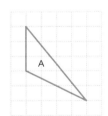

Write down

a the size of angle x

b the length of side y.

3 Give a reason why each pair of triangles is congruent.

a

b

Find the missing angle.

Constructions

To draw and measure a straight line accurately you need to use a ruler.

Here, line *XY* is 5.5 cm or 55 mm long.

The ruler is marked off in cm with the smaller marks showing mm (10 mm = 1 cm).
Line up the 0 on the ruler with the start of the line at *X*.
Measure to the nearest mm and include units in your answer.

Triangles and other 2D shapes

You can use a ruler and protractor to make an accurate drawing of triangle *ABC*.

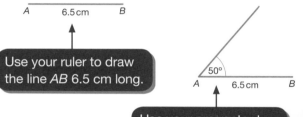

Use your ruler to draw the line *AB* 6.5 cm long.

Use your protractor to draw an angle of 50° at *A* (see page 120).
Draw a long line from *A*.

Use your protractor to draw an angle of 58° at *B*.
Draw a line from *B* so that it crosses the line *AC*.

If you are given the lengths of three sides, you should use a ruler and compasses to construct the triangle, for example *PQR* shown.

Use your ruler to draw the line *PQ* 4 cm long.

Set your compasses to 3 cm.
Put the point on *P* and draw an arc.

Set your compasses to 2.5 cm.
Put the point on *Q* and draw an arc.
The arcs need to be long enough so that they intersect.

Draw a line from *P* to the point where the arcs intersect at *R*.
Repeat with a line from *Q*.

You can now use a ruler, protractor and compasses to make accurate drawings of other 2D shapes including this quadrilateral.

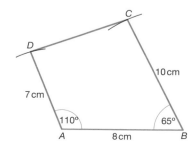

Perpendicular lines

A **perpendicular bisector** is a line that cuts another line exactly in half at right angles.

To construct the perpendicular bisector of the line *AB* you will need a ruler and compasses.

Use your ruler to draw a line to join the points where the arcs intersect.
The line you have drawn is the perpendicular bisector.

Open your compasses to more than half the length of line *AB*.
Put the point on *A* and draw an arc above and below the line.

Keep your compasses set to the same length. Put the point on *B* and draw an arc above and below the line.
The arcs need to be long enough so that they intersect.

You can also use a ruler and compasses to construct the perpendicular to a line that passes through a point *P* on the line.

Other than the first step, the method is the same as for constructing a perpendicular bisector.

Open your compasses. Put the point on *P* and draw arcs on the line either side of point *P*. These points have been labelled *A* and *B*.

Widen your compasses.
Put the point on *A* and draw an arc above and below the line. Then put the point on *B* and draw an arc above and below the line. Use your ruler to draw a line to join the points where the arcs intersect. The perpendicular line you have drawn will go through point *P*.

You can also use a ruler and compasses to construct the perpendicular to a line that passes through a point *P* which is **not** on the line.

Again, other than the first step, the method is the same as for constructing a perpendicular bisector.

Open your compasses. Put the point on *P* and draw arcs on the line either side of point *P*. These points have been labelled *A* and *B*.

Keep your compasses the same.
Put the point on *A* and draw an arc above and below the line. Then put the point on *B* and draw an arc above and below the line. Use your ruler to draw a line to join the points where the arcs intersect. The perpendicular line you have drawn will go through point *P*.

The perpendicular distance from the point *P* to the line is the shortest distance to the line.

DO IT!

Construct the perpendicular bisector of a line *AB*. Use this to draw an isosceles triangle *ABC*.

NAIL IT!

If you are asked to find the **shortest distance** or construct the **shortest path** you should construct the perpendicular line.

Angles

An **angle bisector** is a line that cuts an angle exactly in half.

To construct the bisector of angle *PQR* you will need a ruler and compasses.

Open your compasses. Put the point on *Q* and draw an arc on each arm (*QP*, *QR*) of the angle. These points have been labelled *A* and *B*.	Keep the compasses the same. Put the point on *A* and draw an arc between the two arms of the angle. Then put the point on *B* and draw an arc between the two arms of the angle.	Use your ruler to draw a line from point *Q* to the point where the arcs intersect. The line you have drawn is the angle bisector.

WORKIT!

Use ruler and compasses to construct an angle of 45° at *A*.

Construct the perpendicular bisector of the horizontal line.
Set the compasses to the distance *AB*.
Put the point on *B* and draw an arc on the perpendicular.
Use a ruler to draw a line from point *A* to the point where the arc cuts the perpendicular.

✓ CHECKIT!

In all questions, you must show all your construction lines.

1 Use ruler and compasses to construct an accurate drawing of triangle *ABC*.

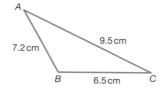

2 Rhombus *ABCD* has sides of length 6 cm. The diagonal *AC* = 5 cm.

Make an accurate drawing of rhombus *ABCD*.

3 Accurately draw the diagram below and use a ruler and compasses to construct the perpendicular to the line segment *AB* that passes through the point *P*.

4 Use ruler and compasses to construct the bisector of angle *ABC*.

5 Use ruler and compasses to construct an angle of 90°.

Drawing circles and parts of circles

You need to know the parts of a circle.

SNAPIT! Parts of a circle

You should know how to draw a circle with a given radius or diameter.

For a circle of radius 4 cm work as shown.

> Mark the centre with a cross (×).
> Set your compasses to 4 cm.
> Put the point of your compasses on the ×
> and draw the circle.

To draw the arc of a circle with a given radius and angle, you will need compasses and a protractor. If you're unsure about using a protractor, see page 119.

DOIT!

Follow these steps to draw an arc of radius 4 cm and angle 50°.

1 Set your compasses to 4 cm.
   Mark the centre with a cross (×) and draw part of the circle.

2 Place the centre of your protractor on the ×.
   Make a mark at 50°.

3 Line up your ruler with the × and the mark.
   Draw a line from the × to the arc.

DOIT!

After completing Question 1 of Check it! write sentences to describe the remaining parts of a circle.

CHECKIT!

1 Complete the following sentences.

   a A _____ is a straight line that does not pass through the centre of a circle but does touch the circumference at each end.

   b A _____ is a straight line that touches the outside of a circle at one point only.

   c A _____ is a straight line through the centre of a circle that touches the circumference at each end.

   d An _____ is part of the circumference of a circle.

   e A _____ is a straight line from the centre to the edge of a circle that is half the length of the diameter.

   f The part of a circle that has a chord and an arc as its boundary is called a _____.

2 Draw a circle of diameter 9 cm.

3 Draw a semicircle of radius 5 cm.

> First work out the radius.

137

Loci

NAIL IT!

A 'set of points' means a straight or curved line (or a region).

A **locus** (plural: **loci**) is a set of points that obey a certain rule.

You need to be able to construct loci using a ruler and compasses (see page 134).

The locus of points that are **equidistant** (same distance) from a point *C* is a circle, centre *C*.

The locus of points that are 3 cm from point *C* is a circle, centre *C*, radius 3 cm.
The shaded region shows the locus of points that are **less than** 3 cm from point *C*.
The region of points that are **greater than** 3 cm from point *C* lies outside the circle.

The locus of points that are the same distance from two points, *A* and *B*, is the perpendicular bisector of the line *AB*.

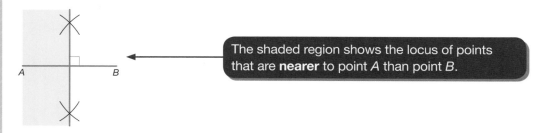

The shaded region shows the locus of points that are **nearer** to point *A* than point *B*.

DO IT!

Lines *AB* and *CD* are parallel and 4 cm apart.

1 Draw lines *AB* and *CD*.

2 Draw the locus of points that are the same distance from line *AB* and line *CD*.

3 Shade the region of points that are nearer to *CD* than *AB*.

The locus of points that are the same distance from two lines, *PQ* and *QR*, is the bisector of the angle *PQR*.

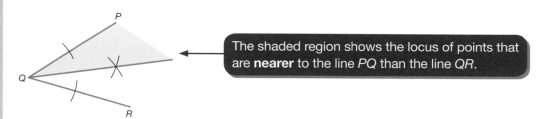

The shaded region shows the locus of points that are **nearer** to the line *PQ* than the line *QR*.

The locus of points that are 2 cm from the line *AB* consists of two lines parallel to *AB* and two semicircles (radius 2 cm).

To solve loci problems you may need to use scale drawings (see page 101) and/or bearings (see page 128).

WORKIT!

The diagram shows a field.

A goat is tethered to a post at *X* by a rope of length 5 m.

Shade the region that can be grazed by the goat.

Scale: 1 cm = 2 m

> The region is part of a circle of radius 5 m from *X*.
> 1 cm represents 2 m, so 2.5 cm represents 5 m.
> Set your compasses to 2.5 cm.
> Put the point on *X* and draw an arc between the two sides of the field.
> Shade the region.

You also need to be able to shade a region that satisfies more than one condition.

In rectangle *ABCD*, the shaded region represents the points that are more than 3 cm from point *C* **and** more than 4 cm from line *AD*.

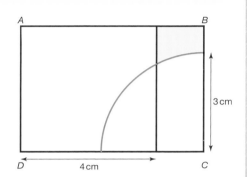

WORKIT!

The diagram shows three radio masts *P*, *Q* and *R*.

Signals from *P* and *Q* have a range of 60 km.

Signals from *R* have a range of 80 km.

Shade the region in which the signals from all three masts can be received.

This diagram is drawn to a scale of 1 cm = 40 km. Yours should be twice as big, with a scale of 1 cm = 20 km.

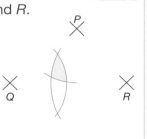

> In your drawing, 1 cm will represent 20 km.
> *P* and *Q*: 3 cm represents 60 km
> Set your compasses to 3 cm.
> Put the point on *P* and draw a large arc. Put the point on *Q* and draw an arc.
> *R*: 4 cm represents 80 km
> Set your compasses to 4 cm.
> Put the point on *R* and draw an arc. Shade the region formed by the arcs.

CHECKIT!

1 Draw the locus of points that are exactly 4 cm from the line *AB*.

2 *RST* is a triangle.

Make an accurate copy of *RST*.

Shade the region inside the triangle that is less than 4 cm from point *S* **and** nearer to the line *RS* than to the line *TS*.

3 Lighthouse *L* is on a bearing of 070° from lighthouse *M*.

Lighthouse *L* is 40 km from *M*.

a Construct a scale drawing to show the positions of lighthouses *L* and *M*. Use a scale of 1 cm = 10 km.

A boat sails so that it is always the same distance from lighthouses *L* and *M*.

b Draw the route taken by the boat.

Perimeter

NAILIT!

Perimeter is a length, so is measured in units of distance – cm, m, km, inches, miles, etc.

The perimeter of a shape is the distance round the outside.

To calculate perimeter, add together the lengths of all the sides.

Sometimes you will need to calculate missing lengths before calculating perimeter.

WORKIT!

Calculate the perimeter of this shape made up of rectangles.

$$\text{Perimeter} = 12 + 9 + 12 + 9$$
$$= 42\ cm$$

The red lines must have a total length of 12 cm since the height of the whole shape is 12 cm. The blue lines must have a total length of 9 cm since the width of the whole shape is 9 cm.

12 cm

9 cm

You can use the properties of shapes to help you find their perimeter.

Shape	Property		Formula
Equilateral triangle	All sides are the same length.	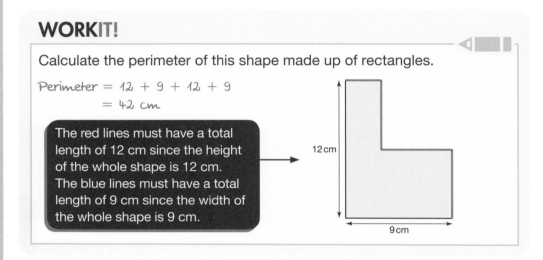	Perimeter = $3a$
Isosceles triangle	Two of the sides are equal in length.		Perimeter = $2a + b$
Square	All four sides have equal length.		Perimeter = $4a$
Rhombus	All four sides have equal length.		Perimeter = $4a$
Rectangle	Two pairs of sides are equal.		Perimeter = $2a + 2b$

Shape	Property		Formula
Parallelogram	Two pairs of sides are equal.		Perimeter $= 2a + 2b$
Kite	Two pairs of sides are equal.		Perimeter $= 2a + 2b$

WORKIT!

Calculate the perimeter of each of these shapes:

a

7.3 cm
4.2 cm
9.1 cm

Perimeter =
9.1 + 7.3 + (2 × 4.2)

    = 24.8 cm

> The dashes marked on the two sloping sides of the trapezium indicate that these two sides are equal.

b

2.8 cm
3.5 cm

Perimeter = (3 × 3.5) + (2 × 2.8)

        = 10.5 + 5.6 = 16.1 cm

> The single dashes on both sides of the triangular top of the shape indicate that these sides are equal. The double dashes on the three sides at the bottom of the shape indicate that all these three sides are equal.

Circumference

The perimeter of a circle is called the **circumference**.

NAILIT!

Circle vocabulary

Circumference = distance round the edge

Diameter = distance across the circle through the centre

Radius = distance from the centre to the circumference

Arc = part of the circumference

Arc
Radius
Diameter
Circumference

SNAPIT!   Circumference

Circumference $= 2\pi r = \pi d$

where r = radius and d = diameter

NAILIT!

The Greek letter Pi (π) represents the ratio of the circumference to the diameter of a circle.

π also represents the area of a circle with a radius of 1 unit.

$\pi = 3.14159265...$

Sometimes you need to give the answer rounded to a number of decimal places or significant figures – if this is the case, use your calculator to work out the answer.

To give the exact answer, leave π in your answer.

NAILIT!

The diameter of a circle is the distance across the circle through the middle. To find the radius you must halve it.

WORKIT!

Calculate the circumference of this circle to 2 significant figures.

$Circumference = \pi \times d$

You are given the diameter, so use the formula $C = \pi d$.

$= \pi \times 14$

$= 43.98229715$

$= 44 \text{ cm}$

14 cm

The perimeter of a semicircle or a quarter-circle can be calculated by using parts of the circumference.

Semicircles

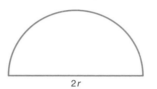
2r

The curved edge has half the length of the circumference of the circle.

Length of curved edge $= \frac{1}{2} \times 2\pi r = \pi r$

Perimeter $= \pi r + 2r$

The straight edge is the diameter, which is double the radius.

Quarter-circles

r

r

The curved edge has one-quarter of the length of the circumference of the circle.

Length of curved edge $= \frac{1}{4} \times 2\pi r = \frac{1}{2}\pi r$

Perimeter $= \frac{1}{2}\pi r + 2r$

Each of the straight edges is a radius.

✓ CHECKIT!

1 Work out the perimeter of a square with sides of length 7.2 cm.

2 Calculate the perimeter of this shape. You may assume all angles are right angles.

7 cm
9 cm
5 cm

3 The perimeter of a semicircle can be calculated using the formula: $k\pi + b$. If the semicircle has radius 4 cm what are the values of k and b?

4 An athletics track has two semicircular ends with radius 15 m. They are connected by straight tracks of 100 m. Find the exact distance around the racetrack.

100 m
15 m

5 A cake is made in the shape of a quarter circle with radius 32 cm. Ribbon is tied around the perimeter of the cake. The ribbon has an overlap of 5 cm.

32 cm

Ribbon costs £0.15 per 10 cm and can only be bought to the nearest 10 cm. Work out the cost of the ribbon for the cake.

Area

The **area** of a shape is the space it takes up in two dimensions.

SNAPIT!   Area formula

Area of rectangle
= base × height
= $b \times h$

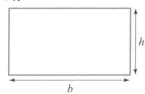

Area of parallelogram
= base × height
= $b \times h$

Area of triangle
= $\frac{1}{2}$ × base × height
= $\frac{1}{2} \times b \times h$

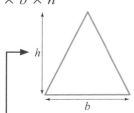

Area of trapezium
= $\frac{1}{2}$ (side a + side b) × height
= $\frac{1}{2}(a + b) \times h$

The height of the triangle is the perpendicular height – that is, how tall it stands. It doesn't have to be the length of one of the sides!

WORKIT!

By how much is the area of a square with perimeter 20 cm larger than that of a triangle with base 8 cm and height 5 cm?

$20 \div 4 = 5$ cm

Area of square = $5 \times 5 = 25$ cm^2

Area of triangle = $\frac{1}{2} \times 8 \times 5 = 20$ cm^2

$25 - 20 = 5$ cm^2 ◄── Subtract to work out the difference.

Work out the length of one of the sides of the square by dividing by 4.

Area of a circle

SNAPIT!   Area of a circle

Area of a circle
= π × radius^2 = πr^2

Radius

The radius of a circle, r, is the distance from the centre of a circle to the perimeter.

WORKIT!

Work out the exact area of a circle with diameter 10 cm.

$10 \div 2 = 5\,cm$ ← First work out the length of the radius.

$\pi \times 5^2 = \pi \times 25$

$= 25\pi\ cm^2$ ← Include π if you need an exact solution.

STRETCHIT!

Write down formulae for calculating the area of:

a a semicircle

b a quarter-circle.

DOIT!

Make some revision cards with pictures of shapes and the formula for finding the area of each shape. Learn them and ask someone to test you.

Composite shapes

A **composite** shape is one that is made up of more than one simple shape.

To find the area of a composite shape, cut it up into shapes with areas you know how to calculate.

For example:

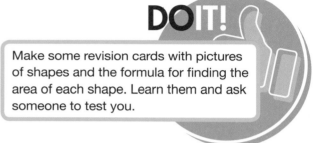

can be split into

Make sure all your measurements are in the same units and find any unknown lengths before you start your calculation.

Sometimes it is easier to calculate a 'missing' area and subtract it from a larger area.

For example, to work out the area of this shape:

you could work out Area A – Area B.

CHECKIT!

1 Work out the area of each of these shapes.

a

2 cm
4.5 cm

b

1.5 cm   2 cm
3 cm

c

5 cm
6 cm   4 cm
9 cm

d

5 cm
7 cm
2 cm

e

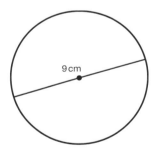

9 cm

Give your answers to 1 decimal place.

2 Work out the area of a square with perimeter 12 cm.

3 Calculate the fraction of this trapezium that is shaded.

6 cm
7 cm
10 cm

4 The rectangles in this shape have equal areas. Work out the total area of the shaded rectangles.

6 cm
8 cm

5 The diagram shows a square enclosing four identical circles.

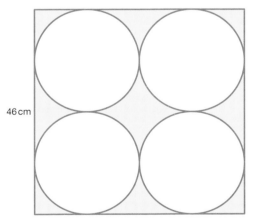

46 cm

Work out the area of the shaded region. Give your answer to 1 decimal place.

Sectors

SNAP IT!

Sector and arc

Area of a sector
$$= \frac{\theta}{360} \times \pi r^2$$

where θ is the angle of the sector.

Length of arc $= \frac{\theta}{360} \times 2\pi r$

A **sector** is a 'slice' of a circle taken from the centre of a circle.

Each slice of pizza is a 'sector'.

Area of a sector

To calculate the area of a sector, work out what fraction of the circle you are calculating and multiply the area of the circle by this fraction.

WORKIT!

Calculate the exact area of the sector.

$$\text{Area} = \frac{60}{360} \times \pi \times 12^2$$

$$= \frac{1}{6} \times \pi \times 144$$

Cancel the fraction down if you can.

$$= 24\pi$$

12 cm

60°

WORKIT!

Work out the perimeter of the sector to 2 decimal places.

4 cm

100°

Length of arc $= \frac{100}{360} \times 2 \times \pi \times 4$

Cancel the fraction down.

$$= \frac{5}{18} \times 2 \times \pi \times 4$$

$$= \frac{20}{9}\pi$$

Perimeter $= 4 + 4 + \frac{20}{9}\pi$

Don't forget to add on the other lengths.

$$= 8 + \frac{20}{9}\pi \text{ cm}$$

$$= 14.98 \text{ cm}$$

Simplify the expression.

Arc length

The **arc** of a circle is part of the circumference.

arc

To calculate the length of an arc, work out what fraction of the circle you are calculating and multiply the circumference by this fraction.

CHECK IT!

1 Work out the area and perimeter of a semicircle with diameter 10 cm.

   Give your answers to 1 decimal place.

2 Calculate the exact area of this sector.

4 cm

3 A semicircular flower bed has a radius of 3 m. Angela buys topsoil to spread over the bed.

   Bags of topsoil cost £14.99 and cover 2 m² each. How much will the topsoil cost? You must show all your calculations.

3D shapes

You need to be able to recognise and name these 3D shapes.

SNAP IT! 3D shapes

Cube Cuboid Cylinder Square-based pyramid

Tetrahedron Triangular prism Sphere Cone

NAIL IT!

A **prism** has the same **cross-section** through its length. This means it has two identical ends. A triangular prism has two (congruent) triangular ends and its cross-section is a triangle. The illustration shows a hexagonal prism – its cross-section is a hexagon.

A **pyramid** has a base and sloping triangular sides that meet at a point. The base could be a square (square-based pyramid), a rectangle (rectangular-based pyramid), a triangle (triangular pyramid or tetrahedron), and so on.

Properties of 3D shapes

Faces, edges and vertices

3D shapes have **faces**, **edges** and **vertices**.

A face is a flat surface of a 3D shape.

An edge is the line where two faces meet.

A **vertex** (plural: vertices) is the point where edges meet.

This tetrahedron has 4 faces, 6 edges and 4 vertices.

Face — Edge

Vertex

STRETCH IT!

Complete the following table.

3D shape	Faces	Edges	Vertices
Cube	6	12	
Cuboid			
Square-based pyramid			
Tetrahedron			
Triangular prism			
Hexagonal prism			

Drawing 3D shapes

You can draw a 3D shape using isometric paper.

You need to make sure that the isometric paper is the right way up! (Vertical lines down the page and no horizontal lines.)

This diagram represents a shape made using four cm cubes.

Nets

A **net** of a 3D shape is a 2D pattern that can be folded up to make the shape.

These are two possible nets for a cube.

You should be able to draw an accurate full-size net of a simple 3D shape.

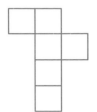

For some 3D shapes you will need to use a ruler and compasses.

Plans and elevations

Plans and **elevations** are 2D drawings of 3D shapes when viewed from different directions.

The side elevation is the view from the side.

The plan is the view from above.

The front elevation is the view from the front.

You need to be able to draw the plan and elevations of a 3D shape.

WORKIT!

The diagram shows a solid made from 6 identical cubes.
On the grid, draw the plan, front and side elevations of the solid.

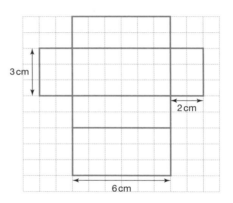

Given the plan and elevations of a 3D shape, you can sketch the shape.

WORKIT!

Here are the plan, front elevation and side elevation of a 3D shape.
Draw a sketch of the 3D shape.

> The two vertical lines on the plan show where the height of the shape changes.

DOIT!

Use cubes to build the 3D shape in the Work it! on page 148. View your shape from above, from the front, and from the side.

Given the plan and front elevation of a 3D shape, you also need to be able to draw the side elevation and then sketch the 3D shape.

CHECKIT!

1 Here is a cuboid.

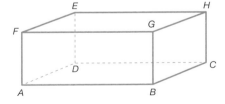

  a Shade in a rectangular face.

  b Shade in the face *BCHG*.

  c Kelli says 'A cuboid has nine edges'. Kelli is wrong. Explain what mistake she has made.

2 The diagram shows a square-based pyramid.

Draw an accurate net of the pyramid. Use a grid of cm squares.

3 Each of these nets will form a 3D shape. Draw a sketch of each 3D shape.

  a   b

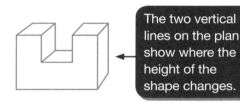

4 The diagram shows a solid made from four identical cubes.
On a grid, draw the plan, front and side elevations.

5 Here are the plan and front elevation of a 3D shape.

  a On a grid, draw the side elevation of the 3D shape.

  b Draw a sketch of the 3D shape.

NAIL IT!

Volume is three-dimensional, so is measured in 'lengths cubed' – for example, cm^3, m^3, km^3, inches^3.

Volume

The volume of a shape is the space it takes up in three dimensions.

SNAP IT!   Volumes of shapes

Volume of cuboid = length × width × height

Volume of sphere = $\frac{4}{3}$ π radius^3 = $\frac{4}{3}$ πr^3

Volume of cylinder = π × radius^2 × height
= $\pi r^2 h$

Volume of a prism
= area of cross-section × length

area of cross-section

length

NAIL IT!

A **pyramid** is any three-dimensional shape that goes up to a point. This is a square-based pyramid, but the formula works for a pyramid with any shape base.

NAIL IT!

A **prism** is a three-dimensional shape with uniform cross-section. This means you could cut a slice and the cross-section would still be the same.

Note that a cylinder is a prism – it has a circular cross-section.

NAIL IT!

A **cone** is a circular-based pyramid.

SNAP IT!   Volume of pyramid and cone

Volume of a pyramid
= $\frac{1}{3}$ × area of base × height

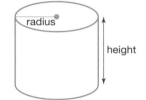

height

area of base

Volume of cone
= $\frac{1}{3}$ × π × radius^2 × height
= $\frac{1}{3}$ π$r^2 h$

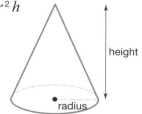

height

radius

Volume of composite shapes

To find the volume of a composite shape, split the shape into two or more 3D shapes with volumes that you can find.

WORKIT!

Workout the volume of this shape.

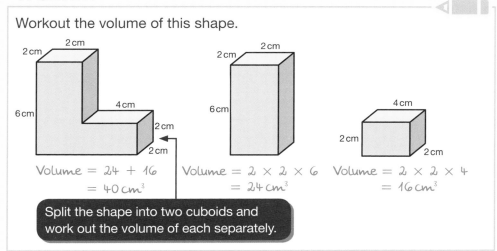

Volume = 24 + 16
      = 40 cm³

Volume = 2 × 2 × 6
      = 24 cm³

Volume = 2 × 2 × 4
      = 16 cm³

Split the shape into two cuboids and work out the volume of each separately.

DOIT!

Make a set of snap cards – on one set of cards draw the 3D shapes, on the other set write the formulae. Shuffle the cards and practise matching the correct formula to the correct shape.

CHECKIT!

1 Calculate the volume of a sphere with radius 4.5 cm. Give your answer to 3 significant figures.

2 A water tank is in the shape of a cylinder on top of a cone. The cylinder is 2 m tall and the whole tank is 3.5 m tall.

If the cylinder has radius 0.5 m, work out the volume of the tank in m³.

3 The volume of this cone is $k\pi$. Work out the value of k.

4 The diagram shows a container partly filled with water to a height of 18 cm.

The container is tipped on its side. What is the height of liquid when the container is on its side?

Surface area

The surface area of a three-dimensional shape is the total area of its faces.

Imagine you have to paint the shape – the surface area is the total area you have to paint.

Calculating surface area can be made simpler by considering faces that have the same area.

For example, this cuboid has pairs of faces with the same area.

Area of top = area of bottom = $9 \times 3 = 27\,\text{cm}^2$

Area of ends = $3 \times 4 = 12\,\text{cm}^2$

Area of front = area of back = $9 \times 4 = 36\,\text{cm}^2$

Surface area = $(2 \times 27) + (2 \times 12) + (2 \times 36)$

$\qquad\qquad = 54 + 24 + 72$

$\qquad\qquad = 150\,\text{cm}^2$

NAILIT!

Surface area is measured in units squared, since you are calculating area.

DOIT!

Find a formula for calculating the surface area of a cuboid measuring a by b by c units.

NAILIT!

Don't just calculate the area of the faces you can see – you need to think about **all** the faces.

You could think about what the net of a shape looks like to help calculate the surface area.

Consider the net of a square-based pyramid.

To calculate the area, first calculate the area of one of the triangular faces and the base.

Surface area = $4 \times$ area of triangle + area of base

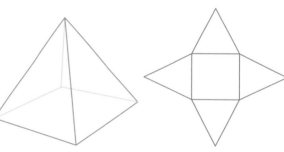

WORKIT!

A glass greenhouse is in the shape of a square-based pyramid with its apex directly above the centre of the base.

What area of glass is used?

Area of triangular side = $\frac{1}{2} \times 4 \times 8$

$\qquad\qquad\qquad\qquad = 16\,\text{m}^2$

Area of glass = 4×16 ← You do not need to add the area of the base in this instance.

$\qquad\qquad\quad = 64\,\text{m}^2$

SNAP IT! Surface areas of shapes

Surface area of a sphere $= 4\pi r^2$

Area of curved surface of a cone $= \pi r l$

WORKIT!

The base of a solid cone has diameter 4 cm and sloping height 8 cm.

Calculate the exact surface area.

Area of sloping face $= \pi \times 2 \times 8$ ← Diameter = 4, radius = 2.

$= 16\pi$

Area of base $= \pi \times 2^2$ ← Don't forget to include the area of the base.

$= 4\pi$

Total surface area $= 16\pi + 4\pi$

$= 20\pi$ cm^2 ← If you are asked for the exact solution, leave your answer in terms of π.

8 cm

4 cm

NAIL IT!

Remember to use the radius when calculating the surface area of a sphere or cone.

CHECKIT!

1 Work out the surface area of a cube with sides of length 5 cm.

2 A sphere has radius 3 cm. Calculate the exact surface area.

3 A shape is made of three 1 cm square cubes which are glued together and painted.

Work out the surface area that is painted.

4 The sloping side of this cone is painted yellow and the base is painted blue.

What percentage of the surface area is yellow?

45 cm

28 cm

Using Pythagoras' theorem

SNAPIT!

Pythagoras' theorem

For a right-angled triangle:

$c^2 = a^2 + b^2$

where c is the **hypotenuse** (the longest side) and a and b are the two shorter sides.

You can use Pythagoras' theorem to find the length of a missing side in a right-angled triangle.

To use Pythagoras' theorem, you need to know the lengths of two sides.

WORKIT!

ABC is a right-angled triangle.

Work out the length of AC.

Give your answer to 2 decimal places.

$c^2 = a^2 + b^2$

$c^2 = 10^2 + 9^2$

$\quad = 100 + 81 = 181$

$c = \sqrt{181} = 13.4536...$

$AC = 13.45\,cm$ (2 d.p.)

Label the hypotenuse c. The hypotenuse is the longest side and is opposite the right angle. Label the other two sides a and b.

Substitute the values of a and b into the formula.

Solve, using a calculator to find the square root.

Round your answer to 2 decimal places and write units with your answer.

The sides could be given as decimal lengths and a range of units (such as mm, cm, m and km) could be used. You also need to be able to give your answer in surd form.

NAILIT!

If you are asked to 'give your answer in surd form' **don't** work out the square root.

WORKIT!

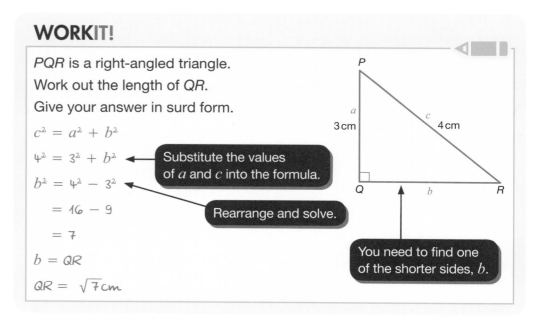

PQR is a right-angled triangle.

Work out the length of QR.

Give your answer in surd form.

$c^2 = a^2 + b^2$

$4^2 = 3^2 + b^2$

$b^2 = 4^2 - 3^2$

$\quad = 16 - 9$

$\quad = 7$

$b = QR$

$QR = \sqrt{7}\,cm$

Substitute the values of a and c into the formula.

Rearrange and solve.

You need to find one of the shorter sides, b.

DOIT!

Show that this triangle is right-angled.

You can show that a triangle is right-angled by showing that $c^2 = a^2 + b^2$.

Solving problems using Pythagoras' theorem

Problems that require the use of Pythagoras' theorem may involve ladders resting against walls, ramps, flagpoles or 2D shapes that contain a right-angled triangle.

WORKIT!

PQRS is a trapezium.

Work out the perimeter of trapezium PQRS.

Give your answer to 1 decimal place.

$c^2 = a^2 + b^2$

$c^2 = 4^2 + 7^2$

$ = 16 + 49 = 65$

> Substitute the values of a and b into the formula for Pythagoras' theorem. Solve for c.

$c = \sqrt{65} = 8.1\,cm$

Perimeter of PQRS

$= 14 + 7 + 10 + 8.1 = 39.1\,cm$

> Use the value of c (PS) to work out the perimeter of PQRS.

> Draw on the line TS to divide the shape into a rectangle TSRQ and right-angled triangle PTS.
> TS = QR = 7 cm
> PT = PQ − SR = 14 − 10 = 4 cm

Length of a line segment

A **line segment** is part of a straight line.

You can use Pythagoras' theorem to find the length of a line segment on a coordinate grid.

WORKIT!

A is the point with coordinates (2, 3)

B is the point with coordinates (5, −1)

Work out the length of AB.

$AB^2 = 4^2 + 3^2$

$ = 16 + 9 = 25$

> Use Pythagoras' theorem to find the length of AB.

$AB = \sqrt{25} = 5$

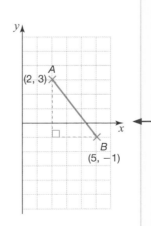

> Draw a right-angled triangle so that AB is the hypotenuse. Work out the length of the two shorter sides:
> Horizontal distance (subtract the x coordinates)
> $= 5 - 2 = 3$
> Vertical distance (subtract the y coordinates)
> $= 3 - (-1) = 4$

NAILIT!

If you're not given a diagram as part of the question, you should first draw a sketch of the two points on a set of x- and y-axes.

CHECKIT!

1 *ABC* is a right-angled triangle.

Work out the length of *BC*.

Give your answer to 3 significant figures.

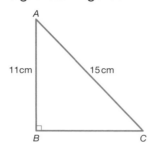

2 A ladder is positioned against a vertical wall at a height of 4.8 m.

The foot of the ladder is 3.6 m from the base of the wall.

How long is the ladder?

3 *XYZ* is a right-angled triangle.

Work out the area of the triangle *XYZ*.

4 Claudia says 'Triangle *PQR* is right-angled'.

Is Claudia correct? Give working and a reason to support your answer.

5 *P* is the point with coordinates (−2, 5)

Q is the point with coordinates (5, −3)

Work out the length of *PQ*.

Give your answer to 2 decimal places.

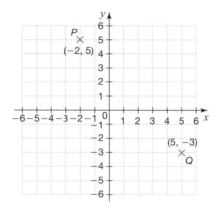

6 *A* has coordinates (2, 0)

B has coordinates (5, 6)

Work out the length of the line segment *AB*.

Give your answer to 1 decimal place.

7 A field *ABCD* is in the shape of a trapezium.

Jesy wants to fence the field.

Fencing costs £14 per metre.

Fencing must be bought in 1 m lengths.

Work out the cost of the fencing.

Trigonometry

Trigonometric ratios are used to calculate the lengths of sides and angles in right-angled triangles.

In order to use the trigonometric ratios you must label the sides of the triangle.

The **hypotenuse** is the side opposite the right angle – it is the longest side.

The **opposite side** is the one opposite the angle you are working with.

The **adjacent side** is the one next to the angle you are working with.

SNAPIT!   Trigonometric formulae

$$\sin x = \frac{\text{opposite}}{\text{hypotenuse}} \qquad \cos x = \frac{\text{adjacent}}{\text{hypotenuse}} \qquad \tan x = \frac{\text{opposite}}{\text{adjacent}}$$

NAILIT!

To help remember the trigonometric formulae, you could learn this phrase.
Studying **O**n **H**oliday   **C**an **A**lways **H**ave   **T**wo **O**bvious **A**dvantages

Calculating the length of missing sides

Here's how to find the length of side *AC* in the triangle shown.

1 Label the sides.

2 Write out SOH CAH TOA and underline the sides you know and the one you want to find out.

   S<u>O</u>H C<u>A</u>H T<u>OA</u>

3 Use the formula that has **two** letters underlined.

   T<u>OA</u>

4 Write out the formula and complete the values you know.

   $\tan 52° = \dfrac{12}{adjacent}$

5 Solve the equation – you may need to rearrange it in order to find the missing length.

   $adjacent = \dfrac{12}{\tan 52°}$

   $adjacent = 9.4\,\text{cm}$

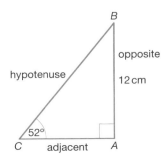

Calculating a missing angle

Here's now to find the size of angle x in this triangle.

1 Label the sides.

2 Write out SOH CAH TOA and underline the sides you know.

S<u>OH</u> CA<u>H</u> T<u>O</u>A

3 Use the formula that has **two** letters underlined.

S<u>OH</u>

4 Write out the formula and complete the values you know.

$$\sin x = \frac{5}{15}$$

5 Find the inverse trigonometric function of your fraction.

$$x = \sin^{-1}\left(\frac{5}{15}\right)$$
$$x = 19.5°$$

DO IT!

Draw a flow diagram to explain how to find a missing side or angle in a right-angled triangle.

NAIL IT!

The inverse trigonometric functions are $\sin^{-1}$, $\tan^{-1}$ and $\cos^{-1}$, which 'undo' the trigonometric functions sin, cos and tan.

Make sure you can find the inverse trigonometric functions on your calculator – usually you have to press 'shift' and then the trigonometric function.

Angles of elevation and depression

An angle measured upwards from horizontal is called an angle of elevation:

angle of elevation

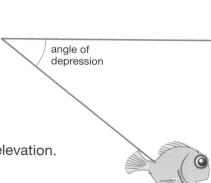

angle of depression

An angle measured downwards from horizontal is called an angle of depression.

The trigonometric ratios can be used to solve problems involving angles of depression and elevation.

WORKIT!

The angle of elevation to a hovering helicopter is 47° and the vertical height of the helicopter from the ground is 100 m. How far away horizontally is the person viewing the helicopter?

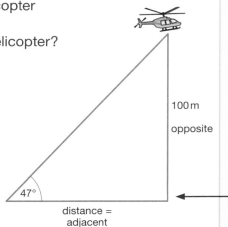

100 m

opposite

SOH CAH TOA

$\tan 47° = \dfrac{100}{\text{horizontal distance}}$

$\text{horizontal distance} = \dfrac{100}{\tan 47}$

$= 93.3\,m$

47°

distance = adjacent

Write out
SOH CAH TOA and underline the sides you know and wish to know, to identify which trigonometric formula to use.

Always draw a picture including all angles and measurements you know.

CHECKIT!

1 Use your calculator to work out each of these values to 1 decimal place.

   a tan 23°

   b sin 35°

   c cos 18°

   d $\tan^{-1} 0.5$

   e $\sin^{-1} \dfrac{3}{4}$

   f $\cos^{-1} \dfrac{1}{\sqrt{3}}$

2 Calculate the length of side *MN*. Give your answer to 2 significant figures.

O

15 cm

72°

M   N

3 Calculate the size of angle *ABC*. Give your answer to 1 decimal place.

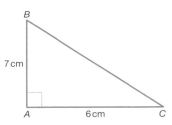

B

7 cm

A   6 cm   C

4 An anchor line from a boat is at 15° to the horizontal. The anchor line is 10 m long.

   How deep is the water below the boat?

15°

10 m

Exact trigonometric values

There are some special trigonometric values which you will need to learn so you can use them in the non-calculator paper.

There are two special triangles that give you the exact trigonometric values.

Triangle 1

Using SOH CAH TOA and considering the 60° angle gives:

$$\sin 60° = \frac{\sqrt{3}}{2} \qquad \cos 60° = \frac{1}{2} \qquad \tan 60° = \sqrt{3}$$

Considering the 30° angle gives:

$$\sin 30° = \frac{1}{2} \qquad \cos 30° = \frac{\sqrt{3}}{2} \qquad \tan 30° = \frac{1}{\sqrt{3}}$$

Triangle 2

Using SOH CAH TOA gives:

$$\sin 45° = \frac{1}{\sqrt{2}} \qquad \cos 45° = \frac{1}{\sqrt{2}} \qquad \tan 45° = 1$$

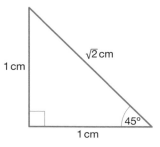

SNAP IT! Trigonometric values

$\sin 60° = \frac{\sqrt{3}}{2}$	$\cos 60° = \frac{1}{2}$	$\tan 60° = \sqrt{3}$
$\sin 30° = \frac{1}{2}$	$\cos 30° = \frac{\sqrt{3}}{2}$	$\tan 30° = \frac{1}{\sqrt{3}}$
$\sin 45° = \frac{1}{\sqrt{2}}$	$\cos 45° = \frac{1}{\sqrt{2}}$	$\tan 45° = 1$

You need to know these too:

$\sin 0° = 0$	$\cos 0° = 1$	$\tan 0° = 0$
$\sin 90° = 1$	$\cos 90° = 0$	

DO IT!

1 Write a ratio fact for each of the trigonometric values in the Snap it!

2 Make a set of cards. On one side write the sin/cos/tan values of 0, 30, 45, 60 or 90.

On the other, write the correct ratio.

You can also write these as ratios.

For example, $\sin 60° = \frac{\sqrt{3}}{2}$ tells you that in a right-angled triangle with an angle of 60°, the ratio between the opposite side and the hypotenuse is $\sqrt{3}:2$.

WORKIT!

Triangle *ABC* has a right angle at *B*.

Angle *BAC* = 60° and side *AC* = 10 cm.

Work out the length of side *AB*.

Label the sides on the triangle.

SOH CAH TOA

$\cos 60° = \dfrac{\text{adjacent}}{\text{hypotenuse}}$

$\dfrac{1}{2} = \dfrac{\text{adjacent}}{10}$

Adjacent = 5 cm

Use the fact that $\cos 60° = \dfrac{1}{2}$

Write SOH CAH TOA and underline the sides you know and want to find out.

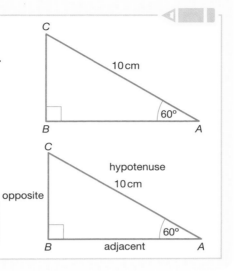

CHECKIT!

Complete these questions without using a calculator.

1 Write down the value of each of these.

a sin 30° **d** sin 45°

b cos 90° **e** tan 60°

c tan 0°

2 Work out the lengths of all the missing sides in this triangle:

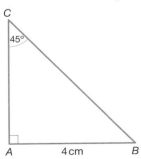

3 In a right-angled triangle, the ratio between the opposite and adjacent sides is 1: $\sqrt{3}$. What are the sizes of the other angles in the triangle?

4 Write down the size of angle *ABC* in the triangle below. Explain how you know.

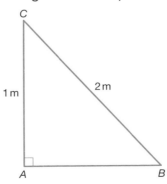

5 Put these values in order of size from smallest to largest:

0.5   cos 30° $\dfrac{3}{4}$ tan 45°

Transformations

Translations

A **translation** is a sliding movement. All the points on a 2D shape move the same distance in the same direction.

To describe a translation you use a **column vector**.

The transformation that moves shape A to shape B is a translation by the vector $\begin{pmatrix} 7 \\ -8 \end{pmatrix}$

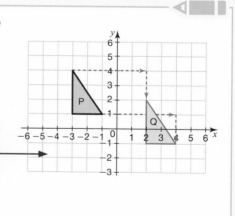

> A positive top number means the movement is to the right.

The top number gives the movement parallel to the horizontal axis (right or left).

The bottom number gives the movement parallel to the vertical axis (up or down).

> A negative bottom number means the movement is down.

The transformation that moves shape A to shape C is a translation by the vector $\begin{pmatrix} -7 \\ 2 \end{pmatrix}$

> A negative top number means the movement is to the left.
> A positive bottom number means the movement is up.

NAILIT!

To describe a translation you need to give the column vector.

DOIT!

Copy the grid and triangle P from Work it! Translate triangle P by the vector

a $\begin{pmatrix} 2 \\ 2 \end{pmatrix}$ **c** $\begin{pmatrix} 0 \\ 3 \end{pmatrix}$

b $\begin{pmatrix} -2 \\ -2 \end{pmatrix}$ **d** $\begin{pmatrix} 3 \\ 0 \end{pmatrix}$

The vector to translate shape A to shape B is $\begin{pmatrix} 5 \\ -1 \end{pmatrix}$

Write down the vector that translates shape B to shape A.

WORKIT!

On the grid, translate triangle P by the vector $\begin{pmatrix} 5 \\ -2 \end{pmatrix}$

Label the new shape Q.

> $\begin{pmatrix} 5 \\ -2 \end{pmatrix}$ means 5 right and 2 down.
> Move each point on triangle P 5 squares right and 2 down.

Translated shapes are **congruent** (see page 132). The lengths of the sides of the shape do not change and the angles of the shape do not change.

Reflections

A **reflection** is obtained when a 2D shape is reflected in a mirror line.

Every point on the **object** (the original 2D shape) and the **image** (the shape following the transformation) must be the same distance away from the mirror line.

You can reflect a shape on a squared grid.

Reflected shapes are congruent.

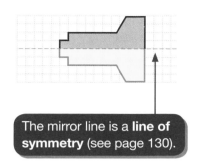

> The mirror line is a **line of symmetry** (see page 130).

WORKIT!

Describe fully the transformation that maps triangle A onto triangle B.

A reflection in the line $y = x$.

> The question says 'describe fully' so you need to write down the type of transformation: a reflection and the equation of the mirror line: $y = x$.

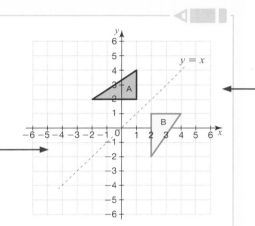

> Draw on the line of symmetry. You can use tracing paper to check that you have identified the correct mirror line. Trace shape A and the mirror line. Remember to mark a reference point on the mirror line, like the origin. Turn the tracing paper over and line up the mirror lines. Does shape A map onto shape B?

NAILIT!

To describe a reflection you need to give the equation of the mirror line.

The mirror line could be:

- a horizontal line: the x-axis ($y = 0$), $y = 1$, $y = 2$…
- a vertical line: the y-axis ($x = 0$), $x = -1$, $x = 1$…
- a diagonal line: $y = x$, $y = -x$

(See page 74 if you're unsure about the equations of straight lines.)

WORKIT!

On the grid, reflect shape P in the line $x = 1$.

Label the new shape Q.

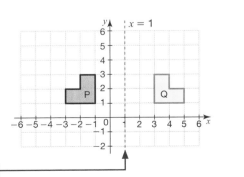

> Draw on the line of symmetry, $x = 1$.

DOIT!

Copy the grid and shape P from Work it! Reflect shape P in

a the y-axis

b the x-axis

c the line $y = 1$

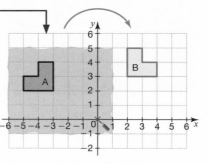

Rotations

A **rotation** can be described as a fraction of a turn or as an angle of turn.

The point about which the 2D shape is turned is called the **centre of rotation**.

Using O as the centre of rotation:

- the rotation of triangle A to triangle B is one quarter turn (90°) clockwise
- the rotation of triangle A to triangle C is one quarter turn (90°) anticlockwise

> When the angle of turn is 180° you don't need to give a direction.

- the rotation of triangle B to triangle C is one half turn (180°).

Rotated shapes are congruent.

WORKIT!

> Mark the centre of rotation with an ×.
> Trace shape A. Put your pencil on the ×.
> Rotate the tracing paper 90° ($\frac{1}{4}$ turn) clockwise.

On the grid, rotate shape A 90° clockwise about the origin, O.

Label the new shape B.

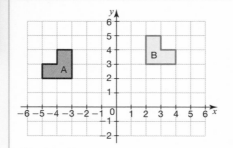

You can also use tracing paper to help you find the centre of rotation.

DOIT!

Copy the grid and shape A from Work it!
Rotate shape A
a 180° about the point O
b 90° clockwise about (1, 0)
c 90° anticlockwise about (0, 2)

WORKIT!

Describe fully the single transformation that maps triangle S onto triangle T.

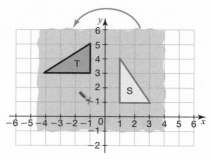

Rotation of 90° anticlockwise about (−1, 1).

> From triangle S to triangle T, the direction is anticlockwise and the turn is 90°.
> Draw lines to find the centre of rotation.
> Or, trace shape S. Put your pencil on a point and rotate the tracing paper. Repeat with a different point until triangle S maps onto triangle T.

NAILIT!

To describe a rotation you need to give:
- the angle of rotation
- the direction of rotation (clockwise or anticlockwise)
- the centre of rotation.

The centre of rotation could be the origin O or any other point on a coordinate grid.

Enlargement

An **enlargement** changes the size of an object. All the side lengths of a 2D shape are multiplied by the same **scale factor**.

Here, shape A has been enlarged by scale factor 2.

> Each length on the enlarged shape is 2 times the corresponding length on shape A.

Enlarged shapes are **not** congruent as the lengths of the sides have changed. Enlarged shapes are **similar** (see page 167).

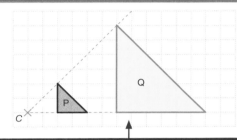

When you are given a **centre of enlargement**, you need to multiply the distance from the centre to each point on the shape by the scale factor.

> For a scale factor of 3, each point on triangle Q is three times as far from the centre C as the corresponding point on triangle P.

WORKIT!

On the grid, enlarge triangle A by scale factor $\frac{1}{2}$ about centre (1, 5).

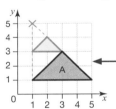

> As the scale factor is less than 1 the image (blue triangle) will be smaller than the object (triangle A). For a scale factor of $\frac{1}{2}$, multiply all the distances from the centre by $\frac{1}{2}$ (or divide all the distances by 2).

WORKIT!

Describe fully the single transformation that maps shape A onto shape B.

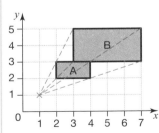

> First work out the scale factor.
>
> Scale factor = $\frac{\text{enlarged length}}{\text{original length}} = \frac{4}{2} = 2$
>
> To find the centre of enlargement, draw lines through corresponding points on the object and image. Extend the lines until they meet at the centre of enlargement.

Enlargement by scale factor 2, centre (1, 1).

DOIT!

On squared paper, draw axes and copy triangle A from Work it! Enlarge triangle A by

a scale factor 2 about centre (1, 1)

b scale factor $1\frac{1}{2}$ about centre (1, 5)

c scale factor 2 about centre (3, 2).

NAILIT!

To describe an enlargement you need to give

- the scale factor **and**
- the centre of enlargement.

CHECK**IT!**

1 On a copy of the grid, draw an enlargement, scale factor 2, of the shaded shape.

2 Describe fully the single transformation that maps shape A onto shape B.

3

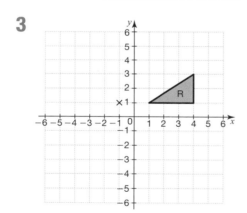

a Rotate triangle R a half turn about the point O. Label the new shape S.

b Rotate triangle R 90° anticlockwise about the point (−1, 1). Label the new shape T.

4 Describe fully the single transformation that maps shape P onto shape Q.

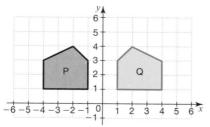

5 Describe fully the single transformation that maps shape A onto shape B.

6

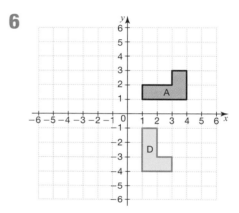

a Reflect shape A in the y-axis. Label the new shape B.

b Translate shape A by vector $\begin{pmatrix} -2 \\ -3 \end{pmatrix}$. Label the new shape C.

c Describe fully the single transformation that maps shape A onto shape D.

Similar shapes

When one shape is an enlargement (see page 165) of another, the shapes are **similar**.

In similar shapes, the corresponding angles are the same and the corresponding sides are in the same ratio. This ratio is the **scale factor** of the enlargement.

Triangles *ABC* and *DEF* are similar.

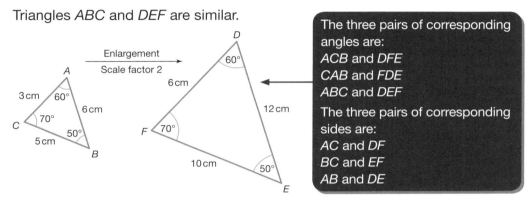

The three pairs of corresponding angles are:
ACB and *DFE*
CAB and *FDE*
ABC and *DEF*

The three pairs of corresponding sides are:
AC and *DF*
BC and *EF*
AB and *DE*

 STRETCHIT!

Work out the perimeter of triangle *ABC* and the perimeter of triangle *DEF*.
When a shape is enlarged by scale factor 2, by what scale factor is the perimeter of the shape enlarged?
Write down a general statement about the perimeter of an enlarged shape.

WORKIT!

Triangle *RST* is similar to triangle *WUV*.

Write down the size of angle *UVW*.

Start by sketching the triangles the same way up.

Angle UVW = angle STR = 30° ◄ Angles in similar shapes are the same.

To work out a corresponding length you first need to work out the scale factor.

Here are two similar triangles, *GHI* and *JKL*.

Take two corresponding sides and work out the scale factor as shown below.

> **Look at two corresponding sides where you know the lengths of both:** *HI* (8 cm) and *KL* (20 cm).

$$\text{Scale factor of enlargement} = \frac{\text{enlarged length}}{\text{original length}} = \frac{20}{8} = \frac{10}{4} = 2.5$$

Once you know the scale factor, you can use this to work out unknown lengths such as *JL* or *GH*.

> **Side *JK* corresponds to side *GH*.**

> **The length of *JL* will be longer than the length of *GI* so multiply by the scale factor.**

Length of *JL* = 9.4 cm × 2.5 = 23.5 cm

Length of *GH* = 12.5 cm ÷ 2.5 = 5 cm

> **Side *GH* corresponds to side *JK*.**
> **The length of *GH* will be shorter than the length of *JK* so divide by the scale factor.**

CHECKIT!

1 Triangle *ABC* is similar to triangle *DEF*.

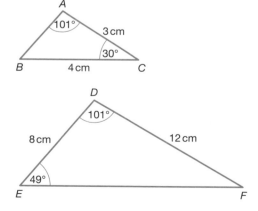

a Write down the size of angle *DFE*.

b Work out the length of *EF*.

c Work out the length of *AB*.

2 Shape *LMNO* is similar to shape *PQRS*.

> **First work out the size of the missing angle in quadrilateral *PQRS*.**

a Write down the size of angle *MLO*.

b Work out the length of *QR*.

c Work out the length of *LO*.

Vectors

A vector has **magnitude** (size or length) and direction.

Vectors can be written as a single bold letter such as **a** or **b**, as a **column vector** (see page 162), or with arrows $\overrightarrow{AB}$, and they can be drawn on squared paper.

You can multiply a vector by a number.

If $\mathbf{a} = \begin{pmatrix} 2 \\ 3 \end{pmatrix}$ then $2\mathbf{a} = \begin{pmatrix} 4 \\ 6 \end{pmatrix}$

$2\mathbf{a} = 2 \times \mathbf{a}$
$= \begin{pmatrix} 2 \times 2 \\ 2 \times 3 \end{pmatrix}$

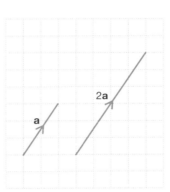

The vectors **a** and **2a** are in the same direction, but **2a** is twice as long.

For a vector **b**, the vector **−b** has the same length but is in the opposite direction.

If $\mathbf{b} = \begin{pmatrix} 3 \\ 1 \end{pmatrix}$ then $-\mathbf{b} = \begin{pmatrix} -3 \\ -1 \end{pmatrix}$ and $-3\mathbf{b} = \begin{pmatrix} -9 \\ -3 \end{pmatrix}$

$-3\mathbf{b} = -3 \times \mathbf{b} = \begin{pmatrix} -3 \times 3 \\ -3 \times 1 \end{pmatrix}$

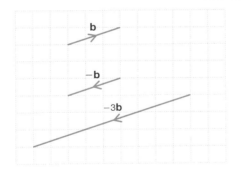

You can add and subtract vectors.

WORKIT!

$\mathbf{a} = \begin{pmatrix} 2 \\ -5 \end{pmatrix}$, $\mathbf{b} = \begin{pmatrix} 3 \\ 4 \end{pmatrix}$

a Work out the vector **a** + **b**.

$\mathbf{a} + \mathbf{b} = \begin{pmatrix} 2 \\ -5 \end{pmatrix} + \begin{pmatrix} 3 \\ 4 \end{pmatrix} = \begin{pmatrix} 5 \\ -1 \end{pmatrix}$

Add the top numbers and add the bottom numbers.

$\begin{pmatrix} 2 \\ -5 \end{pmatrix} + \begin{pmatrix} 3 \\ 4 \end{pmatrix} = \begin{pmatrix} 2 + 3 \\ -5 + 4 \end{pmatrix} = \begin{pmatrix} 5 \\ -1 \end{pmatrix}$

b Work out the vector **b** − 2**a**.

$\mathbf{b} - 2\mathbf{a} = \begin{pmatrix} 3 \\ 4 \end{pmatrix} - 2 \times \begin{pmatrix} 2 \\ -5 \end{pmatrix}$

$= \begin{pmatrix} 3 \\ 4 \end{pmatrix} - \begin{pmatrix} 4 \\ -10 \end{pmatrix} = \begin{pmatrix} -1 \\ 14 \end{pmatrix}$

Subtract the top numbers and subtract the bottom numbers.

Take care with negative numbers: '−' '−' = '+'

DO IT!

This is the vector

$\mathbf{a} = \begin{pmatrix} -2 \\ 2 \end{pmatrix}$

Write **b** as a column vector.

NAIL IT!

Resultant vector

$\overrightarrow{AC} = \overrightarrow{AB} + \overrightarrow{BC}$
$= \mathbf{a} + \mathbf{b}$

The resultant vector of the vectors **a** and **b** is **a** + **b**.

You can draw vectors to represent the sum and difference of two vectors.

The two vectors **a** and **b**	The sum of the two vectors, **a** + **b**	The difference of the two vectors, **a** − **b**:
		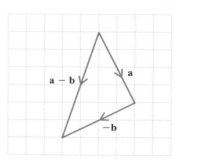

DO IT!

PQRS is a rectangle.

PQ is parallel to *SR*.

SP is parallel to *RQ*.

a Write down $\overrightarrow{SR}$ in terms of **a**.

b Write down $\overrightarrow{RS}$ in terms of **a**.

c Write down $\overrightarrow{QR}$ in terms of **b**.

WORKIT!

$\overrightarrow{BA} = 3\mathbf{a}$, $\overrightarrow{BC} = 2\mathbf{b}$, $\overrightarrow{CD} = \mathbf{b}$

a Write $\overrightarrow{AC}$ in terms of **a** and **b**.

$\overrightarrow{AC} = \overrightarrow{AB} + \overrightarrow{BC}$

$= -3\mathbf{a} + 2\mathbf{b}$ ←

b Write $\overrightarrow{DA}$ in terms of **a** and **b**.

$\overrightarrow{DA} = \overrightarrow{AC} + \overrightarrow{CD}$

$= -3\mathbf{a} + 2\mathbf{b} + \mathbf{b}$

$= -3\mathbf{a} + 3\mathbf{b}$

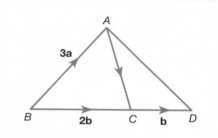

Note the direction of the vectors.

$\overrightarrow{BA} = \mathbf{a}$ so $\overrightarrow{AB} = -\mathbf{a}$

CHECK IT!

1 Draw these vectors on squared paper.

$\mathbf{a} = \begin{pmatrix} 2 \\ 4 \end{pmatrix}$ $\qquad$ $\mathbf{c} = \begin{pmatrix} 3 \\ -2 \end{pmatrix}$

$\mathbf{b} = \begin{pmatrix} -1 \\ 3 \end{pmatrix}$ $\qquad$ $\mathbf{d} = \begin{pmatrix} -4 \\ -6 \end{pmatrix}$

2 Here are two vectors, **a** and **b**.

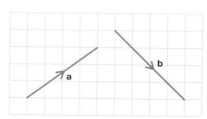

a On squared paper, draw the vector 3**a**.

b Draw the vector **a** − **b**.

c Write −2**a** as a column vector.

3 $\mathbf{a} = \begin{pmatrix} 2 \\ 6 \end{pmatrix}$, $\mathbf{b} = \begin{pmatrix} -5 \\ 3 \end{pmatrix}$

Work out as a column vector

a **a** + **b** $\qquad$ **c** **a** − **b**

b **a** + 2**b** $\qquad$ **d** **b** − 2**a**

4 *PQRS* is a parallelogram.

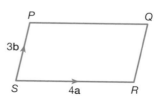

In terms of a and b, work out

a $\overrightarrow{PQ}$ $\qquad$ **c** $\overrightarrow{PR}$

b $\overrightarrow{QR}$ $\qquad$ **d** $\overrightarrow{QS}$

Geometry and measures

You may **not** use a calculator for these questions.

1 Here is trapezium *ABCD*. The diagram is accurately drawn.

a Mark, with the letter *y*, the angle *ABC*.

b Measure the length of line *BC*.

c Measure the size of angle *x*.

2 Here is a triangular prism.

a Write down the number of faces the prism has.

b Write down the number of vertices the prism has.

3 The diagram shows the plan, front elevation and side elevation of a solid shape.

Draw a 3D sketch of the solid shape.

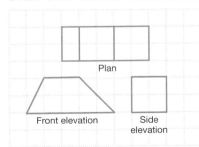

4 On a grid, draw a shape that is congruent to this shape.

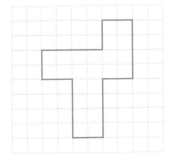

5 Calculate the area of this shape.

6 A square has the same area as this parallelogram. Work out the perimeter of the square.

7 Use a ruler and a pair of compasses to construct the bisector of angle *PQR*. You must show all your construction lines.

8 Describe fully the single transformation that maps triangle A onto triangle B.

9 Work out the size of angle *x*.

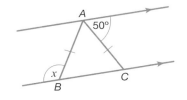

10 What proportion of the rectangle is shaded?

11 Here is triangle *ABC*.

Is triangle *ABC* right-angled?

Give working and a reason to support your answer.

12 Using ruler and protractor only, construct a triangle *PQR* such that *PQ* = 8 cm, angle *RPQ* = 60° and angle *PQR* = 45°.

13 a Write down the exact value of cos 45°.

b Given that $\cos 60° = \frac{1}{2}$, work out the length of side *AB*.

14 $\mathbf{a} = \begin{pmatrix} 4 \\ -5 \end{pmatrix}$, $\mathbf{b} = \begin{pmatrix} 2 \\ 3 \end{pmatrix}$

a Work out vector $\mathbf{a} + 2\mathbf{b}$.

b Work out vector $\mathbf{b} - 2\mathbf{a}$.

You **may** use a calculator for these questions.

15 On squared paper, draw a kite.

16 A shape has three lines of symmetry.

All sides are the same length.

17 Write down the name of the shape.

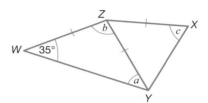

WYZ and *XYZ* are two isosceles triangles with a shared side, *YZ*. *WZ* = *XZ* = *YZ*.

a Work out the size of angle *a*. Give a reason for your answer.

b Work out the size of angle *b*.

c If angle *XZY* = 70°, work out the angle marked *c*.

18 *ABC* is a right-angled triangle.

Work out the length of *BC*. Give your answer to 1 decimal place.

19 The diagram shows part of a pattern made from squares and regular octagons.

Work out the size of angle *x*.

20 *PQR* is a triangle.

Make an accurate copy of *PQR*.

Shade the region inside the triangle that is less than 3 cm from the point *R* and closer to the line *PQ* than the line *QR*.

21 *ABCD* is a trapezium.

Work out the perimeter of trapezium *ABCD*. Give your answer to 2 decimal places.

22 The shape shown is a rectangle with a quarter circle on top.

Calculate the perimeter of the shape to 1 decimal place.

9 cm

4 cm

23 The largest circle possible is drawn inside a square with sides 6 cm long.

Work out the area that is shaded. Give your answer to 1 decimal place.

6 cm

24 The diagram shows a cylindrical glass with height 15 cm and radius 3 cm.

How many times can the glass be completely filled from a 2-litre bottle of water?

15 cm

3 cm

25

5 cm   6 cm
A

11 cm   x
B

Triangles A and B are similar.

Work out x.

26 *ABCD* is a rectangle.

E is the centre of the rectangle.

Work out the length *DE*.

D   A
8 cm
E
C   12 cm   B

27 A solid cone has a circular base with diameter 12 cm, sloping height 10 cm and vertical height 8 cm.

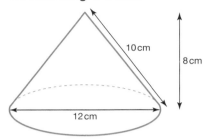

10 cm   8 cm

12 cm

Work out:

a the surface area

b the volume.

Give your answers to 2 significant figures.

28 Calculate the size of angle x in the triangle. Give your answer to one decimal place.

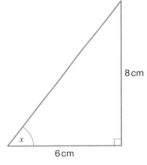

8 cm

x   6 cm

29 Describe fully the single transformation that maps shape A onto shape B.

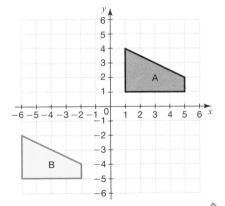

30 Here is triangle *ABC*.

Write down the vector $\overrightarrow{AC}$ in terms of **a** and **b**.

2a + 3b   3a − b

A   C

Probability

Basic probability

The probability of an event is how likely it is to occur.

You can describe an event as:

impossible – it will never happen

unlikely – it probably won't happen

likely – it probably will happen

certain – it will definitely happen

In mathematics, the probability of an event is given a numeric value between 0 and 1. The value is given as a decimal, percentage or fraction.

Probabilities can be marked on a probability scale.

For example, the chance of rolling a dice and it landing on an even number is $\frac{1}{2}$, marked with a cross on the scale.

DOIT!

Write down examples of events that have probabilities of 0, 1 and $\frac{1}{2}$.

SNAPIT!

Common probablities

Learn these common probabilities:

impossible: 0 or 0%

certain: 1 or 100%

even chance, fifty-fifty: $\frac{1}{2}$ or 0.5 or 50%

0        $\frac{1}{2}$ 1

NAILIT!

A probability scale is always between 0 and 1. When marking events on a scale, make sure you divide the line equally – so if you want to mark $\frac{1}{3}$ on the scale, measure the line and divide it equally into three parts.

The **relative frequency** of an event can be used as an estimate for the probability of an event when you cannot calculate it another way.

For example, if you wanted to estimate the probability of a piece of toast landing butter side down you could carry out 100 trials. If it landed butter side down on 60 of those trials, the relative frequency of the event would be:

Relative frequency $= \frac{60}{100} = \frac{3}{5}$

SNAPIT! **Probability and relative frequency**

Probability $= \frac{\text{number of successful outcomes}}{\text{total number of possible outcomes}}$

Relative frequency $= \frac{\text{number of successful outcomes}}{\text{total number of trials}}$

By 'successful' outcomes we mean the relevant ones that we are interested in.

WORKIT!

A bag contains 8 red balls, 1 green ball and 11 blue balls.
Timothy takes one ball from the bag without looking.

a Write down the probability that he takes a green ball. Give your answer as a fraction.

> There is 1 successful outcome since there is 1 green ball.
> There are 20 possible outcomes since there are 20 balls in the bag.

$P(green) = \frac{1}{20}$

b Timothy says there is an 11% chance he takes a blue ball. Is he correct? Explain your answer.

No, he is not correct - the number of blue balls in the bag is more than half the balls, therefore the probability will be greater than $\frac{1}{2}$.

> If you have to explain your answer you should give a mathematical reason or calculation with your answer.

The actual probability of choosing a blue ball is:

$P(blue) = \frac{11}{20}$

> Convert this to a percentage.

$\frac{11}{20} = \frac{55}{100} = 55\%$

NAILIT!

A shorthand for writing 'the probability of getting a green ball' is P(green) or P(g).

Mutually exclusive events

Two events are mutually exclusive if they cannot happen at the same time. For example, rolling a 1 and rolling a 6 on a dice.

If you list all the possible outcomes of an event and they are all mutually exclusive, then the sum of their probabilities is 1.

Look at the spinner. The possible outcomes are red, blue and yellow, which are all mutually exclusive. You can display the outcomes and their probabilities in a table.

Outcomes	Red	Yellow	Blue
Probability	$\frac{1}{6}$	$\frac{1}{2}$	$\frac{1}{3}$

The sum of the probabilities $= \frac{1}{6} + \frac{1}{2} + \frac{1}{3}$

> Convert the fractions so that they all have the same denominator.

$= \frac{1}{6} + \frac{3}{6} + \frac{2}{6}$

> If they all add up to 1, you know you have covered all possibilities.

$= 1$

DOIT!

Write down an example of a pair of events that are mutually exclusive.

WORKIT!

A box contains milk, dark and white chocolates wrapped in foil paper.

The probability of picking a milk chocolate is 0.3.

The probability of picking a dark chocolate is 0.1.

Calculate the probability of choosing a white chocolate.

$P(white) = 1 - (0.3 + 0.1)$

> Since the events are mutually exclusive and the only possible outcomes are choosing a milk, dark or white chocolate, the probabilities sum to 1.

$= 1 - 0.4$

$= 0.6$

Probability of choosing a white chocolate = 0.6

WORKIT!

An unfair spinner is numbered 1 to 4. The table shows the probability of scoring each.

Work out the numeric probabilities.

Outcome	1	2	3	4
Probability	$3p$	$2p + 0.2$	p	$2p$

Simplify the equation by collecting like terms.

$3p + (2p + 0.2) + p + 2p = 1$ ← Since the events are mutually exclusive, the probabilities must sum to 1.

$8p + 0.2 = 1$

$8p = 0.8$

$p = 0.1$ ← Solve the equation to find the value of p.

Substitute the value of p into the expressions.

Outcome	1	2	3	4
Probability	0.3	0.4	0.1	0.2

Probability of an event not occurring

If you know the exact probability of an event occurring you can work out the probability of it not occurring, as they both add up to 1. (It's **certain** that one or the other will happen.)

WORKIT!

The probability that it will rain tomorrow is 0.3. What is the probability of it not raining?

Subtract the probability of it raining from 1.

$1 - 0.3 = 0.7$

SNAPIT!

Probability it won't occur

If the probability of an event occurring is p then the probability of it not occurring is $1 - p$.

Independent and dependent events

If you roll a dice, the outcome you get on the second roll is unaffected by the outcome on the first roll – the probability of scoring each number remains the same.

However, if you are choosing chocolates from a box and eating them, the first choice you make will affect the second choice. You have changed the number of chocolates in the box, so the probability of each outcome changes.

SNAPIT!

Independence and dependence

Events are independent if the outcome of one event does not affect the outcome of the other.

Dependent events are events whose outcomes do affect one another.

WORKIT!

There are 5 green marbles and 2 red marbles in a bag.

a Jenny chooses a marble from the bag. What is the probability that it is red?

$\frac{2}{7}$ ← There are 2 successful outcomes (red marbles) out of a possible 7 outcomes (total marbles).

b Jenny keeps the first marble, then chooses a second marble. Is the probability that it is red still the same? Explain your answer.

No – since she has removed one of the marbles. There are now only 6 marbles in the bag. Now either 1 or 2 red marbles remain in the bag depending on what the colour of the first marble was. The probability will be $\frac{1}{6}$ if the first ball was red, or $\frac{2}{6}$ if the first ball was green.

DOIT!

Write the following terms on a large piece of paper:

- Probability
- Relative frequency
- Mutually exclusive events
- Independent events
- Dependent events

Next to each term, write down all the things you know about it.

CHECKIT!

1 A fair 10-sided dice numbered 1–10 is rolled.

On the number line, mark with an X the probability of rolling a prime number.

2 A packet of sweets contains 12 toffees, 3 boiled sweets and 10 mints. Eileen takes a sweet from the packet at random. What is the probability that she

a takes a boiled sweet

b doesn't take a toffee?

3 Which of the following pairs of events are mutually exclusive?

a Flipping a coin and getting a head
Flipping a coin and getting a tail

b Rolling a dice and getting a prime number
Rolling a dice and getting an odd number

4 A bag of sweets contains red and green sweets.

Which of these pairs of events are independent?

a Picking one sweet at random, eating it, then choosing another

b Picking one sweet at random, replacing it in the bag, then choosing another

5 A 6-sided dice is biased. The probabilities of rolling the numbers 1 to 5 are shown in the table.

Outcome	1	2	3	4	5	6
Probability	0.1	0.15	0.1	0.02	0.2	

Work out the probability of rolling a 6.

6 A spinner has red, green and blue sections. The probability of it landing on green is twice the probability of it landing on red. The probability of landing on blue is 0.4.

What is the probability of landing on green?

Two-way tables and sample space diagrams

NAILIT!

When finding missing values in two-way tables, start by looking for a row or column with only one gap to fill in.

Two-way tables are used when data falls into two different categories.

This table shows the distances (in miles) between three towns in Hampshire:

Petersfield		
12	Waterlooville	
18	9	Portsmouth

The 9 tells you that the distance between Waterlooville and Portsmouth is 9 miles.

This table shows the number of medals won by British competitors in athletics events in an Olympic Games.

The 3 in the Gold column tells you that three gold medals were won by the men. The Total column gives the number of medals won by men and women. The Total row gives the total number of bronze, silver and gold medals.

	Bronze	Silver	Gold	Total
Men	1	0	3	4
Women	0	1	1	2
Total	1	1	4	6

WORKIT!

A shop sells three sizes of T-shirts in four different colours. Each T-shirt can be yellow, red, green or blue and is available in three sizes – small, medium or large.

The two-way table shows some information about the number of T-shirts sold in a single week.

1 Find the sum for the Large row.

2 The first row sums to 12, so 3 green small T-shirts must have been sold. You can now find the total for the Green column.

3 The Red column sums to 15, so 7 medium red T-shirts must have been sold.

4 The Blue column sums to 10, so 3 medium blue T-shirts must have been sold.

	Yellow	Red	Green	Blue	Total
Small	3	1		5	12
Medium			14		34
Large	1	7	2	2	
Total		15		10	

5 The Medium row sums to 34, so 10 yellow medium T-shirts must have been sold. You can now find the sum of the Yellow column.

a How may red T-shirts did the shop sell?

15 ← Look at the total at the bottom of the 'red' column.

b How many medium T-shirts did the shop sell?

34 ← Look at the total at the end of the 'medium' row.

c Complete the two-way table.

6 The sum of the total column and row should be equal, 58. This is a good way to check that you have completed the table correctly.

	Yellow	Red	Green	Blue	Total
Small	3	1	3	5	12
Medium	10	7	14	3	34
Large	1	7	2	2	12
Total	14	15	19	10	58

Sample space diagrams

A sample space (or possibility space) diagram is a type of two-way table that lists all possible outcomes of two events. It can be used to help calculate the theoretical probabilities of outcomes.

NAILIT!

Check if the question uses the word '**and**' or the word '**or**'.
'**And**' means both events must occur.
'**Or**' means one or the other, or both, must occur.

WORKIT!

A fair dice is rolled and a coin is flipped.

a Draw a sample space diagram to show all the possible outcomes.

		Dice					
		1	2	3	4	5	6
Coin	Heads	1, H	2, H	3, H	4, H	5, H	6, H
	Tails	1, T	2, T	3, T	4, T	5, T	6, T

> The cells show the possible outcomes from rolling the dice and flipping the coin.

b Calculate the probability of getting a tail **and** a prime number.

$$P(\text{tail and prime}) = \frac{3}{12}$$

$$= \frac{1}{4}$$

> There are three possible outcomes: 2T, 3T and 5T.

> There are 12 equally likely outcomes in the table.

c Calculate the probability of getting a head on the coin **or** an even number on the dice.

		Dice					
		1	2	3	4	5	6
Coin	Heads	(1, H)	(2, H)	(3, H)	(4, H)	(5, H)	(6, H)
	Tails	1, T	(2, T)	3, T	(4, T)	5, T	(6, T)

> Ring the outcomes with a head or an even number or both.

$$P(\text{head or even}) = \frac{9}{12}$$

$$= \frac{3}{4}$$

> There are 9 equally likely outcomes – 6 with a head and 3 with a tail and an even number.

CHECK IT!

1 Thirty-eight airline passengers are asked for their lunch choices. They may choose chicken, beef or vegetarian option for the main course and fruit or cake for dessert.

Some of the information is shown in the table below.

	Chicken	Beef	Vegetarian
Fruit	12		4
Cake		3	
Total	17	9	

a How many passengers choose the vegetarian option?

b Complete the two-way table.

2 Two fair 6-sided dice numbered 1 to 6 are rolled. The numbers showing on each dice are added together.

A sample space diagram for the possible outcomes is partially drawn below.

		Dice 1					
		1	2	3	4	5	6
Dice 2	1	2	3	4	5	6	7
	2	3	4	5			
	3						
	4						
	5						
	6						

a Complete the sample space diagram.

b Work out the probability that the score is

  i 3

  ii larger than 10

  iii 1

3 A game is played where four cards are placed face down on a table and a player turns over two at random. Each card has the number 1, 2 or 3 on it. The score is found by adding together the two numbers.

The probability of scoring 6 is $\frac{1}{6}$.

The probability of scoring 2 is $\frac{1}{6}$.

What numbers are on the cards?

Sets and Venn diagrams

In mathematics, a **set** is a collection of things.

For example, the set of prime numbers smaller than 20 is:

> The objects in a set are written in curly brackets.

The set of even numbers is written like this:

> These dots tell us the set continues beyond 12.

$\{2, 3, 5, 7, 11, 13, 17, 19\}$ $\{2, 4, 6, 8, 10, 12, ...\}$

The terms in a set are called **elements**. Sets can be represented by **Venn diagrams** (see Snap it! below).

SNAP IT!   Set notation

ξ (Greek letter Xi, pronounced 'Ksi') means the **universal** set – all the elements being considered.

$\in$ means 'is an **element** of' – for example, $3 \in \{\text{prime numbers}\}$.

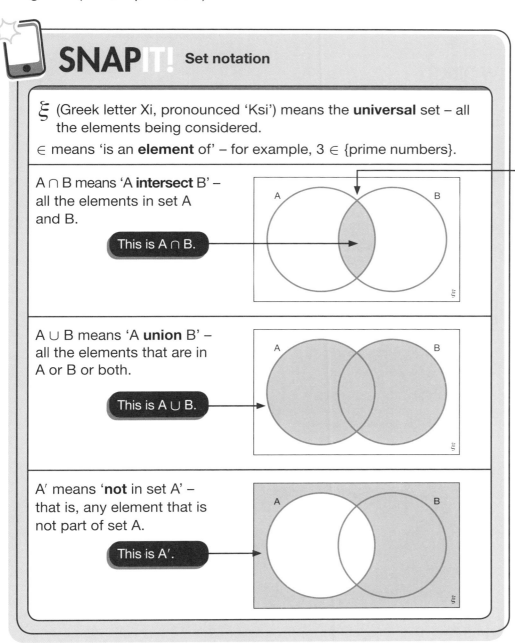

A ∩ B means 'A **intersect** B' – all the elements in set A and B.

This is A ∩ B.

A ∪ B means 'A **union** B' – all the elements that are in A or B or both.

This is A ∪ B.

A′ means '**not** in set A' – that is, any element that is not part of set A.

This is A′.

> A Venn diagram shows each set as a circle containing the elements in that set. The circles can overlap to show that some elements are in more than one set.

Venn diagrams

A Venn diagram is useful because it shows how the elements of more than one set relate to each other.

Venn diagrams can either list all the elements, or give the number of elements in each set.

DO IT!

In example 1, how many elements are there in

a A∪B

b A∩B?

> This symbol means the universal set – all the elements being considered.

Example 1

This Venn diagram shows:

→ ξ = {whole numbers smaller than 20}

> These sets are numbers smaller than 20, so 20 itself is not included.

> Elements in both sets are written in the overlapping area.

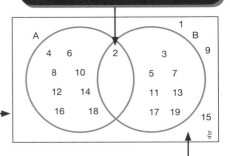

Set A = {even numbers smaller than 20}

Set B = {prime numbers smaller than 20}

In this example, all the elements are listed.

> Elements that do not belong in sets A or B are written outside the circles but are still included since they are in the universal set.

WORKIT!

A group of students is asked whether they play rugby or hockey. R is the set of students who play rugby and H is the set of students who play hockey.

The Venn diagram shows the number of students in each set.

a How many students play neither rugby nor hockey?

17 ← This is the number of students not inside the circles that represent the sets.

b How many students are there in these sets?

> R ∩ H represents all the students who play rugby and hockey.

   i R∩H

    6

   ii R∪H

    15 + 6 + 12 = 33 ← R ∪ H represents all the students who play rugby or hockey or both.

> R′ represents all the students who don't play rugby.

   iii R′

    12 + 17 = 29

   iv ξ

    ξ = 15 + 6 + 12 + 17 = 50 ← This means the universal set – all the students in the survey.

Example 2

This Venn diagram shows the number of people studying French, Spanish and German at A level in Year 12 at a school.

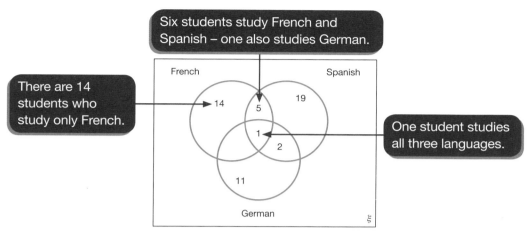

Six students study French and Spanish – one also studies German.

There are 14 students who study only French.

One student studies all three languages.

Calculating probability from Venn diagrams

Venn diagrams can be used to help you work out the numbers you need to use in a probability calculation.

WORKIT!

The Venn diagram shows the number of students in a class with blue eyes and blond hair.

E = {number of students with blue eyes}

H = {number of students with blond hair}

ξ = {students in the class}

A student is selected at random.
Work out the probability that the student

a has blue eyes, P(E)   2 + 4 = 6 so there are 6 students in set E.

$= \frac{6}{30}$ 2 + 4 + 5 + 19 = 30 so there are 30 students all together.

$= \frac{1}{5}$ Don't forget to cancel the fraction down.

b has both blue eyes and blond hair, P(E ∩ H)

$= \frac{4}{30}$ There are 4 students who are in both set E and set H.

$= \frac{2}{15}$

c does not have blue eyes or has blonde hair, P(E′ ∪ H).

$= \frac{28}{30}$ 19 + 4 + 5 are the students **not** in set E and/or in set H.

$= \frac{14}{15}$

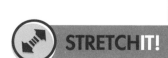

STRETCHIT!

If set A is the set of numbers with two factors and set B is the set of square numbers, how many elements are there in A ∩ B?

NAILIT!

Don't count elements of a set more than once. In the example shown right, to identify the elements in (E′ ∪ H) you need to consider all elements that are **not** in E, and those in H.

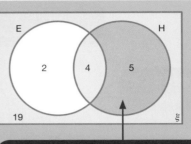

Be careful not to double-count the 5 elements that are just in H.

CHECKIT!

1 ξ = {whole numbers less than 20}

Set A = {multiples of 3 less than 20}

Set B = {multiples of 2 less than 20}

A Venn diagram showing this information has been partially drawn:

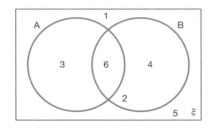

a Complete the Venn diagram.

b Describe the elements of A ∩ B.

2 A year group of 100 students were asked which methods of transport they used to travel to college.

C is the set of students who travel by car.

B is the set of students who travel by bus.

T is the set of students who travel by train.

The Venn diagram shows the number of students in each set.

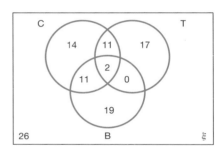

a Complete these sentences. The first is done for you.

C ∪ T is the set of students who travel by car or train or both.

C ∩ T is the set of students who travel by... .

C′ ∩ B is the set of students who... .

b A student from the group is selected at random. Work out

   i P(C)   **ii** P(B ∪ T)   **iii** P(B′ ∩ T)

3 There are 20 elements in sets A and B. Given that P(A) = $\frac{3}{10}$, P(A ∩ B) = $\frac{3}{20}$ copy the Venn diagram and mark how many elements are in each region.

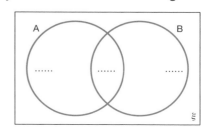

Frequency trees and tree diagrams

DO IT!

Using the frequency tree below, work out the percentage of boys that have school lunch.

Frequency trees

A frequency tree shows two or more events occurring and how many times they happen.

This frequency tree shows the number of boys and girls in a school and whether they have packed or school lunches.

There are 780 girls at the school.

packed lunch — 515

girls — 780

school lunch — 265

1500

packed lunch — 419

boys — 720

school lunch — 301

This is the total number of students at the school.

265 girls have packed lunches.

WORK IT!

A student is randomly selected from the school. Using the frequency tree shown left, work out the probability that the student is a boy who has a packed lunch.

$$\frac{419}{1500}$$

There are 419 boys who have packed lunches.

There are 1500 students at the school.

Tree diagrams

A tree diagram is like a frequency tree, but with the probabilities of the outcomes written on the branches.

Tree diagrams can be used to calculate the probability of events that are both independent and dependent.

The possible outcomes of the first choice are written at the end of the first set of branches.

The possible outcomes of the second choice are then written on the end of each second set of branches.

$\frac{2}{5}$ red

red

$\frac{2}{5}$

$\frac{3}{5}$ blue

$\frac{2}{5}$ red

$\frac{3}{5}$

blue

$\frac{3}{5}$ blue

NAIL IT!

Remember – two events are independent if the result of one does not affect the result of the other.

NAIL IT!

If you add together the probability of all of the possible outcomes you should always get 1, since they are mutually exclusive.

$$\frac{4}{25} + \frac{6}{25} + \frac{6}{25} + \frac{9}{25} = 1$$

Example 1

A bag contains 2 red balls and 3 blue balls. One ball is taken from the bag, its colour is recorded and then it is replaced. A second ball is then taken from the bag and the colour is recorded.

These events are independent since the outcome of the first selection does not affect the outcome of the second selection.

The tree diagram above shows the probabilities of the possible outcomes.

Each of the different 'routes' through the tree diagram gives a different outcome.

You calculate the probability of each 'route' by multiplying the probabilities together.

Red and red $= \frac{2}{5} \times \frac{2}{5} = \frac{4}{25}$ Blue and red $= \frac{3}{5} \times \frac{2}{5} = \frac{6}{25}$

Red and blue $= \frac{2}{5} \times \frac{3}{5} = \frac{6}{25}$ Blue and blue $= \frac{3}{5} \times \frac{3}{5} = \frac{9}{25}$

Example 2

A box of chocolates contains 3 hard centres and 7 soft centres. Maisie chooses a chocolate and eats it; then she chooses a second chocolate.

This tree diagram shows the possible outcomes and the associated probabilities. These events are not independent, since the probability of the second event is affected by the outcome of the first.

> There are only 9 chocolates left in the box when Maisie chooses her second chocolate, so the denominator of the fraction must be 9.

The possible outcomes and probabilities are:

Hard and Hard $= \frac{3}{10} \times \frac{2}{9} = \frac{6}{90}$

Hard and Soft $= \frac{3}{10} \times \frac{7}{9} = \frac{21}{90}$

Soft and Hard $= \frac{7}{10} \times \frac{3}{9} = \frac{21}{90}$

Soft and Soft $= \frac{7}{10} \times \frac{6}{9} = \frac{42}{90}$

NAILIT!

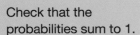

Check that the probabilities sum to 1.

To find the probability of choosing one of each type of chocolate, add together the probabilities of the outcomes 'Hard and Soft' and 'Soft and Hard', since the order in which they are eaten is not specified.

> You must think carefully about how many of each type of chocolate is left in the box when Maisie is making her second choice since she has eaten one chocolate.

$\frac{21}{90} + \frac{21}{90} = \frac{42}{90} = \frac{7}{15}$

WORKIT!

James plays two games at a fair. The probability that he wins the first game is 0.1. The probability that he wins the second game is 0.3.

a Draw a tree diagram to show the possible outcomes.

Game 1   Game 2

0.1 — win
0.3 — win
0.7 — lose

0.9 — lose
0.3 — win
0.7 — lose

> James will definitely either win or lose each game (probability of 1). If the probability that he wins the first game is 0.1 this means the probability he loses this game is 0.9.

b What is the probability that James wins at least one game?

Win and Win = 0.1 × 0.3 = 0.03
Win and lose = 0.1 × 0.7 = 0.07
Lose and Win = 0.9 × 0.3 = 0.27

0.03 + 0.07 + 0.27 = 0.37

> Work out the probability of each way in which James wins at least once. Remember, this includes him winning both games!

> Add together the probabilities to find the total probability.

1 In an orchard there are 347 trees. Of these trees 123 are apple trees and the rest are pear trees.

Of the apple trees 102 are fruiting and of the pear trees 34 are not fruiting.

a Complete the frequency tree to show this information.

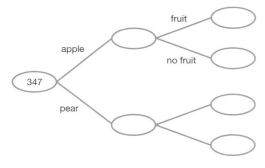

b A tree is chosen at random. What is the probability that the tree is a fruiting pear?

2 A fair 5-sided spinner has the numbers 1 to 5 on it.

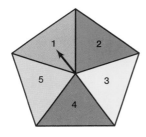

Gary spins the spinner twice and draws a tree diagram showing the possible outcomes.

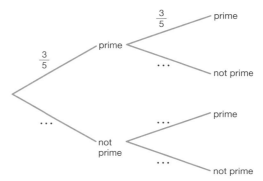

a Complete the diagram.

b Calculate the probability that Gary spins two primes.

c Calculate the probability that Gary spins **at least** one prime.

3 A bag contains 3 glass marbles and 4 plastic marbles. Zac takes one marble from the bag at random, then, without replacing it, he takes a second marble.

a Draw a tree diagram to show the possible outcomes.

b Use the tree diagram to work out the probability that Zac chooses at least one glass marble.

Expected outcomes and experimental probability

Expected outcomes

If you know the theoretical probability of an event occurring then you can calculate how many times it is likely to occur in a given number of trials.

SNAP IT! Number of occurrences

Number of times an event will occur = probability of the event × total number of trials

NAIL IT!

The more times you repeat an experiment, the closer to the theoretical expected number the results will be.

The probability of rolling a 6 on a dice is $\frac{1}{6}$. Therefore, in theory if you roll a dice 12 times you would expect to roll a 6

$\frac{1}{6} \times 12 = 2$ times

In practice this may not actually be the case – this is the **theoretical** probability.

Experimental probability

Sometimes it is necessary to estimate the probability of an outcome using **experimental data**, sometimes called **empirical data**. It is based on observation rather than mathematics or logic.

For example, a student notes the colour of 100 cars passing her house one morning and records the results in a table.

Colour	Frequency
Black	23
Blue	12
Red	18
Yellow	1
White	15
Silver	31

SNAP IT! Experimental probability

Experimental probability = $\frac{\text{frequency of event}}{\text{total frequency}}$

Using the data, she **estimates** the probability of the next car being silver as $\frac{31}{100}$.

The experimental probability can be used to estimate how many silver cars you would expect to find in a larger sample.

In a sample of 1000 cars, you would expect $\frac{31}{100} \times 1000 = 310$ cars to be silver.

SNAPIT! Using empirical data

Here are some things you need to remember when using empirical data.

1. The larger the sample, the more accurate the experimental probability.

2. Experimental probability is only an estimate since only a sample has been used.

3. Just because one outcome hasn't occurred doesn't mean it isn't possible.

Bias

If something is biased then one thing is more likely to happen than another – it is 'unfair'.

For example, if a bag contains coloured balls, but the red ones are twice the size of the other colours, then even if you were trying to pick out a ball at random you would be more likely to pick out the red balls. The experiment would be biased.

When conducting probability experiments you should make sure that the experiment is not biased by ensuring every outcome has an equally likely chance.

DOIT!

Make a short list of real-life situations where you have to use experimental probability rather than mathematical probability.

WORKIT!

A biased dice is rolled 20 times and the outcomes are shown in the table below:

Outcome	1	2	3	4	5	6
Frequency	0	2	11	4	1	2

a What is the experimental probability of rolling a 3?

$\frac{11}{20}$ ← The number of times a 3 was rolled is 11. The total number of times the dice was rolled is 20.

b Does the fact that a 1 was not rolled mean that it is impossible to roll a 1? Give a reason for your answer.

Not necessarily. Although a 1 was not rolled, this does not mean it is impossible – it just might not have been rolled yet.

c Jenny rolls the dice 100 times. How many times can she expect to roll a 2?

$\frac{2}{20} \times 100 = 10$ ← Multiply the experimental probability by the number of times the dice is rolled.

STRETCHIT!

A dice is rolled 12 times and none of the rolls give a 6. Is this enough rolls to say that the dice is biased? Explain your answer.

CHECKIT!

1 The probability of passing your driving test is 0.45. In one month, 300 people sit their driving test. How many people would you expect to pass?

2 A factory produces bags of sweets. Each bag contains 100 sweets. A sample of 10 sweets is chosen from the bag and their colour recorded.

Colour	Frequency
Red	2
Green	6
Yellow	1
Orange	1

Find an estimate for the number of sweets in the bag that are red.

3 A dice is rolled 100 times. If the dice is fair, how many prime numbers would you expect to roll?

4 Three friends carry out an experiment to discover if buttered toast is more likely to land butter side down. Their results are as follows:

	Charlie	Nicolai	James
Butter side down	112	10	28
Butter side up	74	7	19

a Whose results will give the best estimate for the probability of the toast landing butter side down? Justify your answer.

b Charlie carries out the experiment a further 10 times. Use the combined results of the experiment to work out how many times he can expect the toast to land butter side down.

You may **not** use a calculator for these questions.

1 Approximately 12% of people are left-handed. In a school of 250 students how many would you expect to be left-handed?

2 A fair dice is rolled. Mark the probability of each of these events on a probability scale.

A – rolling an even number

B – rolling a factor of 6

C – rolling a 7

D – rolling a number smaller than 8.

3 The probability that it will rain tomorrow is 0.3. What is the probability that it won't rain?

4 Which of the following pairs of events are mutually exclusive:

A – rolling a prime number and rolling an even number on a dice

B – rolling a 3 and rolling a factor of 7 on a dice

C – rolling a 1 and rolling an even number on a dice.

5 A fair 5-sided spinner is labelled with the letters A to E.

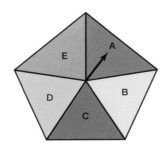

a What is the probability of **not** spinning a vowel (A or E)?

b The spinner is spun 25 times. How many times can you expect to spin a letter B?

6 Complete the two-way table showing lunch choices:

	Pizza	Pasta	Risotto	Total
Cake	12			
Ice cream		11	10	
Total	22	17		50

7 A box of chocolates contains milk, white and dark chocolates.

The table shows the probability of picking each type of chocolate.

Chocolate	Milk	White	Dark
Probability	0.2	$5x + 0.2$	x

Work out the probability of picking a white chocolate.

8 Yan and Ethan have two dice. They think that the dice might be biased.

Yan rolls his dice four times and gets: 2, 2, 6, 1.

a Yan says 'my dice must be biased'. Is he correct? Justify your answer.

b Ethan rolls his dice 50 times and records his results.

Score	1	2	3	4	5	6
Frequency	12	9	16	7	6	0

Ethan decides to roll his dice another 100 times.

Using the results in the table, estimate the number of times he should expect to roll a 2.

9 During Holly's journey to work she cycles through two sets of traffic lights. The probability of a red light at the first set of lights is 0.1. The probability of a red light at the second set of lights is 0.5.

The tree diagram shows all the possible outcomes of her journey.

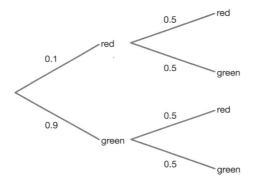

Work out the probability that at least one of the sets of lights is red.

10 Aidyn flips a coin and rolls a dice. If the coin shows heads he doubles the score on the dice. If it shows tails he adds 2 to the score on the dice.

a Complete the possibility space diagram for each possible score.

		Dice					
		1	2	3	4	5	6
Coin	Head	2	4				
	Tail	3	4				

b Find the probability of

i a score of 4

ii scoring less than 4.

11 Here is a Venn diagram:

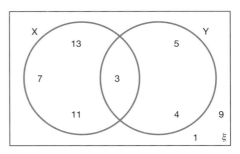

a How many elements are there in

i X ∪ Y **ii** X ∩ Y **iii** Y′?

b A number from the set is chosen at random.

Work out the probability that the number is a member of X.

12 A bag contains 2 red balls and 7 green balls.

Alice takes a ball from the bag at random and keeps it, then takes another ball from the bag.

a The tree diagram shows the outcomes and probabilities. Complete the tree diagram.

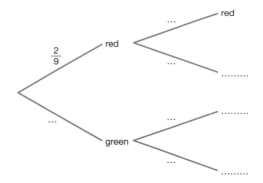

b Use the tree diagram to work out the probability that Alice takes 2 green balls from the bag.

13 A computer randomly generates fractions using the digits 1 to 4 – for example, $\frac{3}{4}$. The numerator is always smaller than the denominator.

a It generates 30 fractions. How many of the fractions can you expect to be less than $\frac{1}{2}$?

b Give a reason why the number of fractions generated that are less than $\frac{1}{2}$ may not be exactly the same as your answer to part a.

14 In a primary school of 300 students 45% are boys.

$\frac{4}{5}$ of the girls and $\frac{2}{3}$ of the boys play sports after school.

A child is picked at random. What is the probability the child plays sport after school?

15 A fairground game has 20 floating ducks on a pond. Players hook out a duck and if the duck has a cross on the bottom it is a winner. There are five ducks with crosses on the bottom.

The game costs £1 to play. If you hook a duck with a cross you win £2.

If 100 people play the game, how much profit should the fairground game make?

You **may** use a calculator for this question.

16 Two students are carrying out a survey to try to work out what percentage of the population is left-handed. Their results are shown in the table below.

	Erica		Milo	
	Male	Female	Male	Female
Right-handed	23	18	51	60
Left-handed	5	4	7	7

a Whose results will give the better estimate for the probability of someone being left-handed? Justify your answer.

b Use the combined results to work out the number of left-handed students you would expect to find in a school with 2000 students.

Statistics

Data and sampling

Statistical investigations involve considering the following questions before collecting any data:

- What problem needs solving?
- What information is needed?
- How should information be collected?
- How should information be presented and displayed?
- How should the information be interpreted?
- What conclusions can be drawn?

SNAP IT! Quantitative or qualitative?

Data can be:

- **quantitative** – numerical values (for example, age)
- **qualitative** – describing a quality of a thing (for example, colour)

Data that is collected by the person using it is known as **primary** data. For example, if you asked people in the street what their favourite chocolate bar is, the data would be primary.

Data collected by someone else is known as **secondary** data. For example, if you looked up sales of chocolate on the internet, the data would be secondary.

When carrying out statistical research about a population you may need to take a sample, as it is cheaper and quicker to take a sample than it is to ask the whole population.

For example, if you wished to find out what the most popular chocolate bar in the country is, then the population would be everyone in the country. It would take too long and cost too much to survey everyone in the country so you would take a sample.

SNAP IT!

Primary data – you ask the questions yourself.

Secondary data – someone else asks the questions.

In a **random sample** every member of the population has an equal chance of being chosen.

A **biased** sample is one where a certain type of person or thing is more likely to be chosen. For example, if you wished to find the nation's favourite chocolate bar and you asked all the people in your class, the sample would not be a good representation of the whole population. It would consist of people of the same age who all live in the same area.

SNAP IT! Population or sample?

The **population** is the complete set of things you are interested in.

A **sample** is a small part of the population.

To ensure the sample represents everyone in the country you would need to ensure that you include people:

- of different ages
- of different social and economic backgrounds
- who live in different areas
- of different cultural backgrounds.

You must also make sure the sample is large enough to be a good representation of the population, but not so large that it is too costly or time consuming.

 STRETCHIT!

> If you wanted to find out what the people in your school enjoyed doing in their free time, how would you choose your sample?

WORKIT!

A scientist wishes to investigate the life expectancy of emperor penguins. The population of emperor penguins is approximately 600 000. The scientist takes a sample of 600 birds, all aged between 2 and 3 years old.

Give two reasons why this is not a good sample.

The sample size is small.

She is only sampling $\frac{600}{600\,000} \times 100 = 0.1\%$ of the population.

The sample is biased since she is only considering birds who are between 2 and 3 years old, therefore every member of the population does not have an equal chance of being chosen.

 NAILIT!

Try to include mathematical values when commenting on sample size.

 STRETCHIT!

> If the population you are considering is split into distinct groups, one way of sampling is to ensure the number of people from each group that you sample is proportional to the size of the group.
>
> This is called **stratified sampling**.
>
> For example, if there are twice as many German people as there are French people at a meeting and you took a stratified sample, you should sample twice as many German people as French people.
>
> To work out how many of each group to include in a stratified sample:
>
> **1** Calculate the percentage or fraction of the population that is in that group.
>
> **2** Multiply the sample size by the percentage or fraction.

You will not be asked questions about stratified sampling in the exam, but it's a useful technique to know about!

Using sampling to describe a population

Once you have collected data using a sample, it can be used to predict the characteristics of the whole population.

For example, a mobile phone company surveys 6000 mobile phone users and finds that 1500 are using their network.

It uses this information to predict that $\frac{1500}{6000} \times 100 = 25\%$ of the population of mobile phone users are using their network.

Since there are approximately 42.4 million mobile phone users in the UK, the company predicts that 25% of 42.4 million = 10.6 million people use their network.

NAIL IT!

The larger the sample, the more accurate the predictions that can be made about the population.

✓ CHECK IT!

1 A class of students wishes to find out about the average temperature in the UK in November. Give one primary and one secondary source of data that they may use.

2 A television company carries out a survey about favourite TV channels. It asks viewers to tell them which channel they watch most often. What type of data is this?

3 Give two reasons why taking a sample is a good way of collecting statistical data.

4 An animal charity carries out a survey on how people feel about wearing real fur. They ask all the people who work for their charity.

  a Explain why this sample would be considered biased.

  b Describe how to choose a sample that is unbiased.

5 A car manufacturing company makes 800 000 cars a year. A sample of 200 cars is checked on the same date every year to ensure that there are not any faults in the production. The company discovers that 3 of the 200 cars have a fault.

  a How many faulty cars can the company expect to have produced in a year?

  b Give two reasons why your answer to part a might be unreliable.

Frequency tables

SNAP IT! Discrete and continuous data

There are two different types of quantitative data:

Discrete data – this can only take particular values – for example, shoe sizes can be 1 or $1\frac{1}{2}$, but not $1\frac{3}{4}$.

Continuous data – this can take any value within a range – for example, your height can be measured to any degree of accuracy, such as 1.4357 m.

Grouped data

Frequency tables are a good way of recording data.

This frequency table shows the number of A level subjects taken by a group of students:

Number of A level subjects	Frequency
3	9
4	6
5	5
Total:	20

You can work out the total number of students by adding together the frequencies:

Total number of students
$= 9 + 6 + 5 = 20$

To work out the total number of A levels studied, multiply each number of A level subjects by its frequency, then find the sum.

$3 \times 9 = 27$ ← 9 students studied 3 A levels.

$4 \times 6 = 24$ ← 6 students studied 4 A levels.

$5 \times 5 = 25$ ← 5 students studied 5 A levels.

Total A levels studied $= 27 + 24 + 25 = 76$

SNAP IT! Designing a frequency table

When designing frequency tables for grouped data follow these rules:

1. Aim for between 4 and 6 groups.

2. The groups should have equal width.

3. The groups must not overlap or have gaps between them.

NAILIT!

- When grouping data, be careful the groups you are using don't overlap.

  For example, if you choose groups of:
  0–1 minute
  1–2 minutes
  for times taken to complete a task, in which group would you record somebody who took exactly 1 minute?

- Although grouping data is useful because it enables you to display large amounts of data in a small space, the original values are not recorded and so some of the accuracy of the data is lost.

$45 \leq a < 60$ means that customer's age is higher than or equal to 45 but lower than 60.

First work out how wide the classes will be. Decide whether to divide them into 4, 5 or 6 groups by dividing the total width by 4, 5 and 6.

Discrete data

A railway company wishes to record the number of passengers in each of its train coaches at a particular time of day. The data can only take whole number values (people). This is known as **discrete** data. You can write groups like 1–5, 6–10 and so on.

The frequency table might look like this:

Number of passengers	Frequency
1–25	
26–50	
51–75	
76–100	

The frequency column tells you how many coaches they recorded with a number of passengers in the given range.

Continuous data

A manager of a gym wishes to collect information about the age of customers. The data being collected (age) is continuous so inequalities must be used with no gaps between them.

Age in years (a)	Frequency
$15 \leq a < 30$	
$30 \leq a < 45$	
$45 \leq a < 60$	
$60 \leq a < 75$	

NAILIT!

To determine whether to use $<$ or $\leq$, decide if you **want** that value included in that class – if you do, use $\leq$.

Each class should have one $<$ and one $\leq$.

WORKIT!

A researcher wishes to find out about the average time teenagers spend on their homework each week.

She predicts they will spend between 0 and 25 hours a week on homework.

Design a data collection sheet for her data.

$25 \div 4 = 6.25$
$25 \div 5 = 5$
$25 \div 6 = 4.17$

Choose a sensible class width that is easy to work with.

Width of class $= 25 \div 5 = 5$ hours

Total number of hours (h)	Frequency
$0 \leq h < 5$	
$5 \leq h < 10$	
$10 \leq h < 15$	
$15 \leq h < 20$	
$20 \leq h < 25$	

Teenagers may do no homework at all, so you must use the $\leq$ sign at the beginning of the class.

CHECKIT!

1 A bus driver records the number of people on her bus at 20 different times during the day. Her results are as follows:

3, 18, 9, 27, 14, 19, 10, 11, 11, 4, 19, 33, 18, 24, 15, 17, 17, 20, 7, 12

Design and complete a grouped frequency table for the data.

2 a A gardener records how many courgettes each of his 15 plants produced. None of his plants produced more than 5 courgettes.

Use this information to complete the frequency table.

Number of courgettes	Frequency
0	1
1	0
2	1
3	1
4	9
5	
6	

b How many courgettes did he grow altogether?

3 A football coach records the hours each of his team spend training each week. He draws up the following table to record the information.

a Give two mistakes he has made.

b The coach says, 'the longest time anyone in the team spends training is 40 hours'. Explain why he might be wrong.

Total number of hours (h)	Frequency
$12 \leq h < 15$	
$16 \leq h < 19$	
$20 \leq h < 30$	
$30 \leq h < 40$	

Bar charts and pictograms

Bar charts

A bar chart can be used to represent discrete data.

For example, the data about the number of A levels studied by a group of students can be displayed in a bar chart.

To work out the total number of students, add together the heights of the bars.

Total number of students
$= 9 + 4 + 5 = 18$

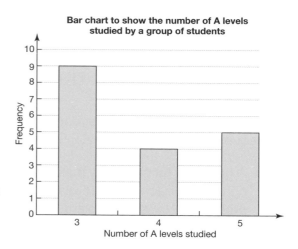

Bar chart to show the number of A levels studied by a group of students

Bar line charts

These are exactly the same as bar charts – but with lines drawn instead of bars.

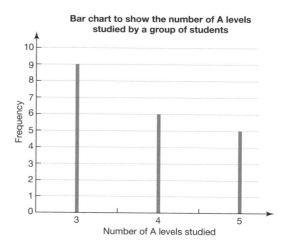

Bar chart to show the number of A levels studied by a group of students

Comparative bar charts

A comparative bar chart can be used to easily compare sets of data.

From this comparative bar chart, you can see at a glance that there are more boys in both the maths and physics classes.

The key is essential

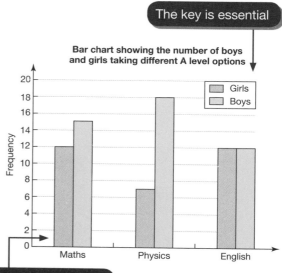

Bar chart showing the number of boys and girls taking different A level options

12 girls study maths
15 boys study maths

DOIT!

From the comparative bar chart on the right, how many more boys study physics than girls?

Composite bar charts

This composite bar chart displays the same information as the comparative bar chart. It is easier to see at a glance the total number of students in each lesson, but more difficult to identify the exact number of boys or girls in each.

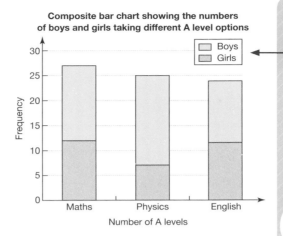

Composite bar chart showing the numbers of boys and girls taking different A level options

Remember to include a key.

Bar charts for grouped discrete data

A bar chart can also be used to display grouped discrete data.

This bar chart shows the number of students in evening classes offered by a further education college.

The total number of classes can be found by adding together the heights of the bars:

total number of classes
= 9 + 11 + 1 = 21

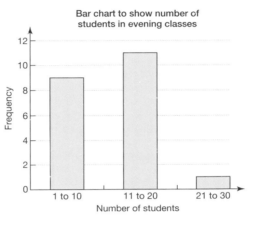

Bar chart to show number of students in evening classes

Pictograms

Pictograms can be used to display discrete data.

For example, here is a pictogram displaying numbers of A levels that students take.

The key states each symbol represents 2 students, so the total number of students in the sample can be calculated by counting the symbols and multiplying by 2.

Number of A levels	Number of students
3	
4	
5	

= 2 students

CHECKIT!

1 A group of students were asked how many pets they owned.

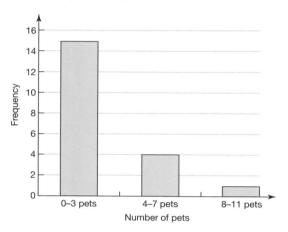

a How many students were surveyed?

b What percentage of students owned four or more pets?

2 A group of college students were asked how many sports they played regularly. The results are shown in the compound bar chart.

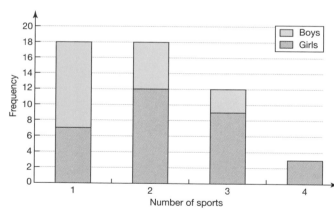

a How many more boys than girls played only one sport?

b What percentage of students surveyed were boys?

c Jasmine says that the proportion of boys to girls who play two sports is larger than the proportion of boys to girls who play three sports. Show that this is true.

3 A shop records the size of shoes it sells in one day. The bar line chart displays some of the information.

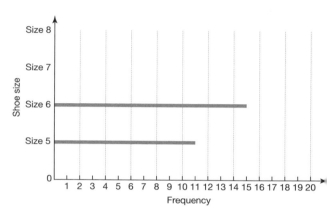

The shop sells a total of 50 shoes on the day. It sells twice as many size 7 shoes as size 8 shoes. Copy and complete the chart.

4 The pictogram shows the colours of bikes manufactured at a factory.

a How many more black bikes than red bikes were made?

b What proportion of the bikes made were silver?

Pie charts

Drawing pie charts

SNAPIT!   Drawing a pie chart

To draw a pie chart, follow these steps:

1. Work out the total number of items your pie chart represents.

2. Divide 360° by the total number of items. ◄—— *This gives the size of a pie slice per item.*

3. Multiply the angle (your answer to Step 2) by the number of items in each group – this gives you the angle for each group.

4. Construct the pie chart – use a protractor to measure the angles.

5. Label the regions or include a key.

WORKIT!

The favourite sport of a group of students is recorded in this frequency table. Calculate the angles of a pie chart to represent the data.

Favourite sport	Frequency
Football	22
Rugby	15
Tennis	12
Swimming	11

Total number of students = 22 + 15 + 12 + 11

= 60 ◄—— Find the sum of the frequencies.

360° ÷ 60 = 6° ◄—— Divide 360° by the sum of the frequencies. 6° represents one student.

Football = 22 × 6° = 132°

Rugby = 15 × 6° = 90°

Tennis = 12 × 6° = 72° ◄—— Calculate the angle needed for each sport.

Swimming = 11 × 6° = 66°

NAILIT!

Check the angles sum to 360° – if they don't, you have made a mistake somewhere!

For this example,
132° + 90°
+ 72° + 66° = 360°

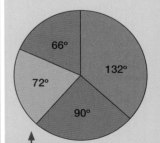

You would not actually write the angles on your chart.

Football
Rugby
Tennis
Swimming

Interpreting pie charts

The pie chart shows the distribution of different flavours of 30 sweets in a packet.

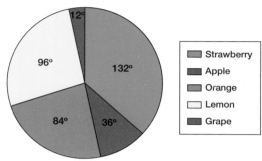

You can work out how many sweets there are of each flavour by working out how many degrees represent each sweet.

$360° \div 30 = 12°$ ← *Divide 360° by the number of sweets. 12° represents each sweet.*

Strawberry $= 132 \div 12 = 11$ sweets

Apple $= 36 \div 12 = 3$ sweets

Orange $= 84 \div 12 = 7$ sweets

Lemon $= 96 \div 12 = 8$ sweets

Grape $= 12 \div 12 = 1$ sweet

$11 + 3 + 7 + 8 + 1 = 30$ sweets ← *Check your answer – the total number of sweets should be 30.*

Divide each angle by 12 to work out how many sweets each section represents.

DOIT!

Draw a flow chart to describe how to construct a pie chart.

STRETCHIT!

If the angles are not whole numbers, how would you deal with that?

WORKIT!

The pie chart shows the ages (x) of 120 people who use a swimming pool.

a How many people are between 18 and 40?

20% are in the group

$18 < x \leq 40$

10% of $120 = 12$

$20\% = 2 \times 12 = 24$ people.

b How many more people using the pool are under 18 than are between 18 and 40?

70% of $120 = 7 \times 12$

$\qquad = 84$

$84 - 24 = 60$ more people ← *Subtract the number of people between 18 and 40 from the number of people under 18.*

CHECKIT!

1 Students at a school choose to study French, Spanish or German.

Draw a pie chart to represent this data.

Language	Number of students
French	27
Spanish	42
German	21

2 The pie chart shows the annual salaries of 240 students leaving university.

a How many students earned over £40 000 a year?

b What proportion of students earned less than £30 000?

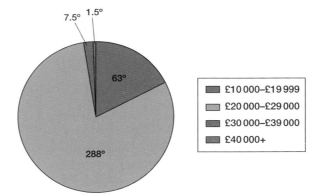

3 Stefan uses a pie chart and a bar chart to display data he has collected about the number of cars owned per household.

Number of Cars	Frequency	Angles
0	6	18
1	21	63
2	65	195
3	18	54
4	10	30

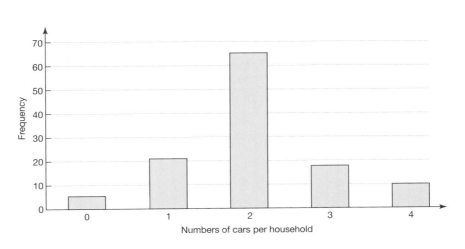

a How many households owned 3 or more cars?

b Which chart is it easier to read this from? Explain why.

Measures of central tendency: mode

'Measures of central tendency' are often referred to as 'averages'. There are several different types of measures of central tendency.

Mode or modal class

The **mode** is the value that occurs most often in a set of data.

If the data is grouped, the **modal class** is the one with the highest frequency.

The mode is the only measure of central tendency (average) that can be found from both quantitative and qualitative data, it is easy to find and can show what the most 'popular' value or thing is.

Finding the mode from a list of data

The colour of 10 cars is recorded, with results as follows:

blue, green, white, silver, silver, blue, red, silver, white, black

Which is the modal colour?

The mode is silver since this occurs three times – more than any other colour.

Sometimes there is no mode since no data value is repeated more than once or each value is repeated the same number of times.

Sometimes there is more than one mode, but we do not usually give more than two modes, instead we say there is no mode.

The age, in years, of swimmers in a race was recorded:
11, 11, 13, 11, 13, 12, 12, 11, 13, 12, 13, 12, 11, 13

The modal values are 11 and 13. This is not a particularly useful piece of information.

Finding the mode from a frequency table

If the data is displayed in a frequency table, the mode or modal group is the one with the highest frequency.

For example, this table shows the ages of students taking GCSE maths at a school.

Age of students taking GCSE maths	Frequency
15	32
16	144
17	19

The modal age of students taking GCSE maths is 16, since this is the age of the largest number of students.

The second table shows the ages of students at a school.

The modal class is $11 < a \le 12$ since this is the class with the largest number of students.

Age of students (a)	Frequency
$10 < a \le 11$	112
$11 < a \le 12$	125
$12 < a \le 13$	110
$13 < a \le 14$	109
$14 < a \le 15$	100
$15 < a$	92

NAILIT!

The 'modal' value is the one that occurs most often – it means the same as mode!

Finding the mode from a bar chart

It is easy to identify the mode from a bar chart. It is the tallest bar.

Here the mode is maths since this is the subject that the greatest number of students study.

Finding the mode from a pie chart

The modal value is the one represented by the largest sector.

In the survey of students' starting salaries from page 205, the modal class is £20 000–£29 999, since this is the largest segment.

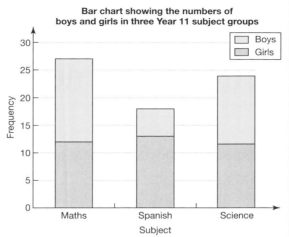

Bar chart showing the numbers of boys and girls in three Year 11 subject groups

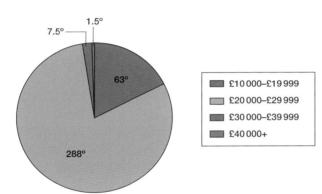

CHECK IT!

1 Five numbers have a mode of 12.2. Given that two of the numbers are 11.9, what can you say about the other numbers?

2 A group of students records the length of time they spend on the internet in a day.

What is the modal class?

Time, t (hours)	Frequency
$0 \leq t < 1$	13
$1 \leq t < 2$	62
$2 \leq t < 3$	12
$3 \leq t < 4$	5
$4 \leq t < 5$	8

3 Alice has recorded the number of pets each member of her class has.

Number of pets	Frequency
0	0
1	7
2	14
3	8
4	0
5	0

Alice says, 'the modal number of pets is 0 since 0 appears three times in the frequency column'.

Max says that Alice is wrong.

Who is correct? Explain why.

Measures of central tendency: median

The median is the middle value of an ordered set of data.

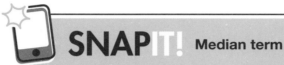

SNAPIT! **Median term**

If there are n pieces of data, the median term is the $\frac{n + 1}{2}$ th term.

NAILIT!

You must make sure the items of data are ordered before finding the median.

WORKIT!

Find the median of these values: 2, 3, 1, 4, 4, 6, 9, 2, 7

1 2 2 3 ④ 4 6 7 9

Median term $= \frac{9 + 1}{2} = 5^{th}$

There are nine terms in the list. The median is the 5th term.

The median is 4.

Finding the median from a frequency table

If data is grouped, you must first work out the total number of pieces of data.

The table shows the number of siblings a group of students have.

Number of siblings	Frequency
0	3
1	5
2	7
3	1
Total:	16

If it isn't provided, first find the sum of the frequencies.

NAILIT!

If the median term lies between two values, add together the values and divide by 2.

Median term $= \frac{16 + 1}{2} = 8.5$th term

The median term lies between the 8th and 9th terms.

You could list all the terms:

0   0   0   1   1   1   1   1   2   2   2   2   2   2   2   3

The 8th and 9th terms are 1 and 2, therefore the median is $\frac{1 + 2}{2} = 1.5$.

Finding the median of grouped data

Finding the actual value of the median from grouped data is beyond the scope of this revision guide, but you can find the median class.

WORKIT!

The length of time a group of people spend exercising each week is recorded. Which is the median class?

Time, t (hours)	Frequency
$0 \leq t < 2$	7
$2 \leq t < 4$	15
$4 \leq t < 6$	18
$6 \leq t < 8$	3

$7 + 15 + 18 + 3 = 43$ ← Find the sum of the frequencies.

Median term $= \dfrac{43 + 1}{2} = 22$nd term

The 22nd term is in the class $2 \leq t < 4$

Median class $= 2 \leq t < 4$

✓ CHECKIT!

1 Find the median of the list of numbers below

3.1   9.2   8.7   6.5   4.3   2.9

2 A group of students records the number of wild birds they see in a week.

What is the median class?

Birds, b	Frequency
$0 \leq b < 2$	29
$2 \leq b < 4$	28
$4 \leq b < 6$	30
$6 \leq b < 8$	3
$8 \leq b < 10$	10

3 The median of a list of numbers is 3.5. Given that all the numbers are integers explain why there cannot be five numbers in the list.

Measures of central tendency: mean

When we talk about the average we are usually talking about the mean.

SNAPIT! Mean

$$\text{Mean} = \frac{\text{sum of all data values}}{\text{total number of values}}$$

WORKIT!

Calculate the mean of 10, 15, 19, 7 and 4.

$$\text{Mean} = \frac{10 + 15 + 19 + 7 + 4}{5}$$

Divide by 5 since there are five data values.

$$= \frac{55}{5}$$

$$= 11$$

Calculating the mean from a frequency table

Siblings are brothers and sisters.

This table shows the number of siblings that a group of students have.

In order to work out the mean number of siblings you need to calculate:

total number of siblings ÷ total number of students

Number of siblings	Frequency
0	3
1	19
2	13
3	1

Draw a third column on the table:

Number of siblings	Frequency	Number of siblings × frequency
0	3	0 × 3 = 0
1	19	1 × 19 = 19
2	13	2 × 13 = 26
3	1	3 × 1 = 3
Total:	36	48

3 people had 0 siblings.

19 people had 1 sibling.

The total number of siblings.

The total number of students.

$$\text{Mean} = \frac{48}{36} = 1\frac{1}{3} \text{ siblings}$$

STRETCHIT!

Is this a sensible answer? You know the answer must lie between 0 and 3 siblings. But if you wanted a whole number you might choose to use the mode instead.

Calculating the mean from a bar chart

This bar chart shows the number of pets owned by students in a class.

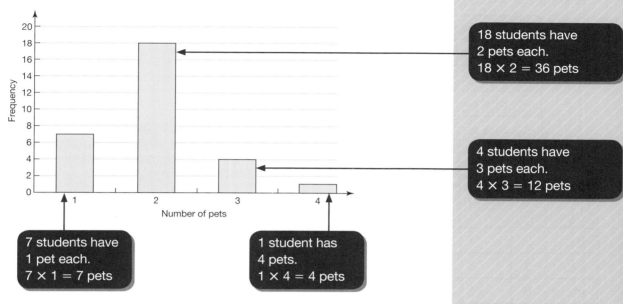

18 students have
2 pets each.
18 × 2 = 36 pets

4 students have
3 pets each.
4 × 3 = 12 pets

7 students have
1 pet each.
7 × 1 = 7 pets

1 student has
4 pets.
1 × 4 = 4 pets

To work out the mean number of pets, work out:

total number of pets ÷ total number of students

The total number of students is the sum of the heights of the bars:

7 + 18 + 4 + 1 = 30

To work out the total number of pets, multiply the height of each bar by the number of pets it represents:

Total number of pets = (1 × 7) + (2 × 18) + (4 × 3) + (1 × 4)

$$= 7 + 36 + 12 + 4 = 59$$

Mean number of pets $= \dfrac{59}{30}$

$$= 2.0 \text{ pets (to 1 decimal place)}$$

 STRETCHIT!

Which measure of central tendency would you use in each of the following situations?

a Average shoe size.

b Average length of time spent waiting for a bus.

c Average amount of money spent buying a new car.

Estimating the mean of grouped data

You cannot find the exact mean from grouped data since you do not know what the values of the original data were. However, you can find an estimate for the mean.

Use the midpoint of the class as the data value.

WORKIT!

The length of time a group of people spend exercising each week is recorded.

Time, t (hours)	Frequency
$0 \leq t < 2$	7
$2 \leq t < 4$	15
$4 \leq t < 6$	18
$6 \leq t < 8$	3

Calculate an estimate for the mean to one decimal place.

> Add another column to calculate the midpoint of the class. Use this as an estimate of the actual time spent exercising.

Time, t (hours)	Midpoint	Frequency	Frequency × midpoint
$0 \leq t < 2$	1	7	$1 \times 7 = 7$
$2 \leq t < 4$	3	15	$3 \times 15 = 45$
$4 \leq t < 6$	5	18	$5 \times 18 = 90$
$6 \leq t < 8$	7	3	$7 \times 3 = 21$
	Total:	43	163

> 7 people spend 1 hour exercising.

> The total time spent exercising.

Mean $= \frac{163}{43} = 3.8$ hours

> The total number of people surveyed.

CHECKIT!

1 Andrea records the length of time she spends waiting for her bus in one month.

Time, t (mins)	Frequency
$0 \leq t < 4$	12
$4 \leq t < 8$	3
$8 \leq t < 12$	5

a Calculate an estimate for the mean length of time Andrea waits each day.

b Explain why your answer is an estimate.

2 In a quiz the mean score of five students is 9. Another student scores 6 marks.

What is the new mean?

3 The mean of three numbers is 5. One of the numbers is 5.

Yasmina says that the other numbers must also be 5.

Is she correct? Explain your answer.

Range

The range is a measure of how spread out data is.

WORKIT!

Calculate the range of the numbers 21, 67, 91, 15, 19, 18.

Range = 91 − 15

= 76

SNAPIT!   Range

Range = largest value − smallest value

Finding the range from a bar chart

Here's the bar chart again showing the number of pets that students own.

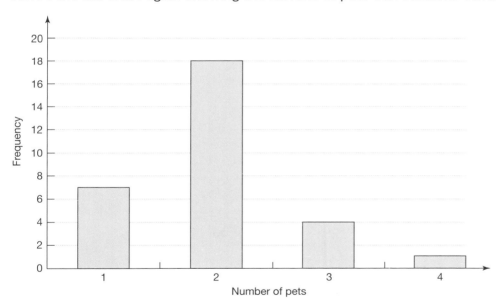

To find the range, subtract the smallest number of pets from the largest number:

Range = 4 − 1 = 3

NAILIT!

Don't make the mistake of subtracting the frequencies – these are **not** the data values.

CHECKIT!

1 Calculate the range of the following values
3.1, 4.2, 9.5, 8.7, 5.2, 9.1, 9.2, 4.9, 0.7, 1.4

2 The bar chart show the age of students studying A level German at a school.

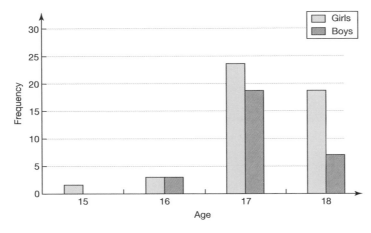

Work out the range of ages of the

a girls   **b** boys.

3 The times in seconds taken by two athletes to run 100 m are shown in the table below.

Athlete A	Athlete B
13.2	14.3
14.5	14.8
13.0	14.5
13.9	15.2
14.7	14.9
15.2	15.0

Which athlete has the greater range of times?

4 In a science test the range of marks is 20. The lowest mark is 45%.

Another student takes the test and the range increases to 30.

What are the possible marks the student could have achieved?

Comparing data using measures of central tendency and range

Measures of central tendency and range are useful as they enable you to compare data.

Here is the data on the distances, to the nearest metre, that two sportsmen throw a javelin.

	Mean length of throw	Modal length of throw	Median length of throw	Range
Sportsman A	81 m	82 m	82 m	16 m
Sportsman B	81 m	98 m	79 m	38 m

Which sportsman would you choose for your team?

Either answer can be correct as long as you justify your answer using the data.

You might choose Sportsman A since his range is lower, therefore he is more consistent in the length of his throws.

You might choose Sportsman B since his modal throw is much higher than Sportsman A's.

DO IT!

Make flash cards with 'Mean', 'Median' and 'Mode' on them and write down the advantages and disadvantages of each.

Advantages and disadvantages of each measure of central tendency

	Advantages	Disadvantages
Mode	1 Quick and easy to calculate. 2 The mode of qualitative data can be found. 3 You can find the mode or modal class of discrete and continuous data. 4 Extreme values do not affect the mode.	1 It does not represent the whole data set since extreme values are ignored. 2 There may be more than one mode. 3 There may be no mode – for example, if each value occurs only once.
Median	1 Extreme values do not affect the median.	1 It can be time-consuming to find for large data sets. 2 It does not take into account any extreme values.
Mean	1 All the data is used to find the answer.	1 Extreme values can distort the data. 2 It can be time-consuming to calculate.

Your choice of measure of central tendency can affect the conclusions you draw from a statistical investigation.

WORKIT!

A battery manufacturer wishes to show how long-lasting its batteries are.

A sample of five batteries is tested and the length of time they last for, in the same device, is recorded in hours. The results are as follows:

15.4, 18.9, 2.4, 14.7, 19.6

a Which average would the company use in its advertising? Explain your answer.

$$Mean = \frac{15.4 + 18.9 + 2.4 + 14.7 + 19.6}{5} = 14.2$$

Median:

2.4, 14.7, 15.4, 18.9, 19.6 ← Write the data in order first.

Median = 15.4

Mode: There is no mode

The company would use the median since this is the highest value.

b A rival company uses the range of the data to show how unreliable the batteries are. Explain why they use this calculation.

The range is 19.6 − 2.4 = 17.2

This shows the lifespan of the batteries can vary greatly.

NAILIT!

If asked to explain your answer, use calculations to support your decision if possible.

CHECKIT!

1 The length of time children spend on a computer game over one weekend is recorded:

32 mins, 29 mins, 18 mins, 41 mins, 362 mins, 19 mins

Work out:

a **i** the mean   **ii** the median.

b Why are the averages so different?

2 A school uses the mean to calculate the average age of students in the year. Give a reason why they might have chosen to use this measure of central tendency.

3 A car manufacturer records the times it takes two different cars to accelerate from 0 to 60 miles per hour.

They carry out several experiments. The results are shown below.

Car A: Mean = 5.3 seconds
     Range = 0.9 seconds

Car B: Mean = 4.2 seconds
     Range = 2.3 seconds

Which car would you recommend? Justify your answer.

4 A children's shoe shop records the size of wellington boots sold one morning.

Shoe size	Pairs of boots sold
3	2
4	5
5	18
6	2

a Which average would be a useful measure of central tendency?

b Explain why you have chosen this average.

Time series graphs

A time series graph is used to show how data changes over time and can be used to spot trends over time.

This graph shows how the temperature changes during a single day.

Take care when reading particular values from a time series graph. Remember that the line joining two points does not show accurate values between those points.

Plotting a time series graph

SNAPIT!   Plotting a time series graph

When plotting a time series graph, remember to:
- plot the time along the x-axis (the horizontal axis)
- keep the intervals regular along both axes
- join the points with straight lines
- label the axes.

DOIT!

Use this graph to work out:

a the highest temperature

b the lowest temperature

c the range of temperatures.

WORKIT!

The height of a tomato plant is measured over 20 days.

a What height were the seedlings on day 11?

6 cm ◄ Draw a vertical line from the x-axis at day 11, and a horizontal line from the point the vertical line touches the time series line, and read off the value.

b What height would you expect the seedling to be on day 21? Explain whether you think your answer is accurate.

16 cm. The prediction is unreliable since it is outside the scope of the data values recorded.

217

Constructing a table for a time series

To construct a table to display a time series, make sure that:

- the time intervals are regular
- all the relevant times are included.

For example, if you want to record air temperature over a 24-hour period, you might produce a table like this one.

> The 24-hour period has been split into 12 equal intervals.
>
> You could split it into more or fewer intervals as long as they are the same length.

Time	0	2	4	6	8	10	12	14	16	18	20	22	24
Temperature													

CHECKIT!

1 The table shows the speed of an accelerating car over a 1-minute interval.

Draw a time series graph to represent the data.

Time (s)	0	10	20	30	40	50	60
Speed (m/s)	0	15	24	32	40	45	47

2 The graph shows the rate at which a cup of tea cools over time.

a What temperature is the tea after 15 minutes?

b What temperature would you expect the tea to be 60 minutes after the experiment begins?

c Is your prediction reliable? Explain your answer.

3 The graph shows the number of tourists who visit a holiday destination each year.

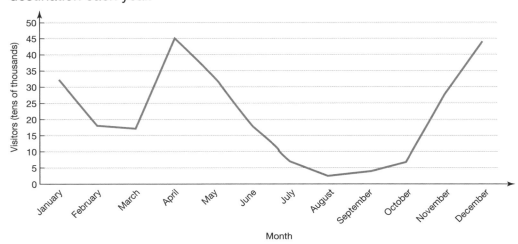

a How many tourists visited in March?

b During which month were there: i the most tourists   ii the fewest tourists?

c Describe the pattern in the number of tourists over the year.

Scatter graphs

A scatter graph is a good way to identify relationships between two sets of data.

The table shows the results of 10 students in their science and maths tests.

Maths	73	92	34	19	35	34	74	70	38	82
Science	69	83	50	20	29	100	69	70	44	88

These results are plotted in a scatter diagram:

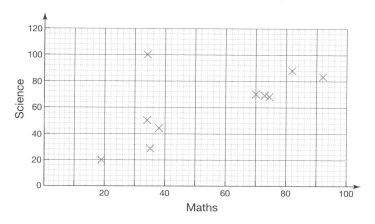

The relationship between the two sets of data is called **correlation**.

NAILIT!

If you are asked to describe the relationship between two sets of values in a scatter diagram, you need to decide if it is positive, negative or if there is no correlation. You must also explain what this means.

SNAP IT! Correlation

Positive correlation
As the x value increases, the y value increases.

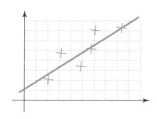

Negative correlation
As the x value increases, the y value decreases.

No correlation
There is no relationship between the x and y values.

In the above example, the data for the students has a positive correlation. This means that as the maths mark increases, the science mark increases.

NAILIT!

Correlation does not imply causation! Just because there is correlation between two things it does not mean one causes the other.
For example, people on a lower income are statistically more likely to be smokers – but just because they are on a lower income does not **cause** them to become smokers.

Line of best fit

A line of best fit can be drawn on a scatter graph if there is positive or negative correlation.

Draw a straight line following the trend of the data – you should aim to have the same number of points on each side – but ignore any outliers.

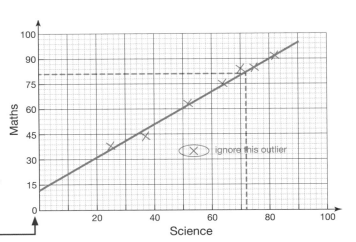

Mark your axes carefully making good use of the graph paper you have. Would there have been a more helpful way of marking these axes?

SNAPIT! **Outliers**

An **outlier** is a value that doesn't fit with the trend.

The outlier is ignored – this value may be caused by a student being unwell on the day of the test or other influences, so it should not be allowed to affect the trend of the data.

The line of best fit can be used to predict missing values.

For example, you could predict the mark that a student who scored 72% in the maths exam is likely to achieve in science. Draw a vertical line up to the graph from 72 on the x-axis, then read across horizontally to the y-axis – giving 81%.

Reading data values between the lowest and highest values of a data set is called **interpolation** – it is quite reliable because it is within the recorded data.

In the example above, you can confidently predict that the student's maths mark is likely to be around 81%.

NAILIT!

The line of best fit does **not** have to go through the origin.

Taking a data value from outside the data set is called **extrapolation** – it is less reliable since you do not know that the data will continue to follow this trend beyond the highest and lowest values recorded.

If the student had scored 15% in science, this value is lower than the lowest science score – so reading the maths mark from the graph will be less reliable. It doesn't mean you can't do this – but any answers found using extrapolation should be treated with caution!

NAILIT!

Remember any values you read from the line of best fit will only be estimates since your line of best fit may be slightly different from those drawn by other people.

SNAPIT! **Interpolation or extrapolation?**

Interpolation
Values read from within the data set – these are fairly reliable

Extrapolation
Values read from outside the data set – these are less reliable

WORKIT!

The scatter graph shows the average daily temperature plotted against the number of heaters sold by a department store.

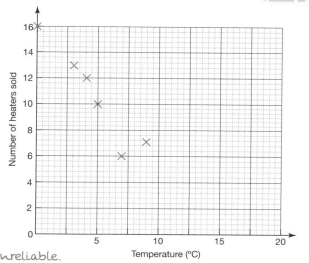

a Describe the type of correlation and explain what this means.

Negative correlation – the higher the average temperature, the fewer heaters are sold.

b Explain why you would not use a line of best fit to predict how many heaters you would sell on a day when the average temperature was 15°C.

15°C lies outside the range of the data so you would be extrapolating, therefore the prediction would be unreliable.

CHECKIT!

1 The scatter diagram shows the length and weight of seven babies born in a hospital.

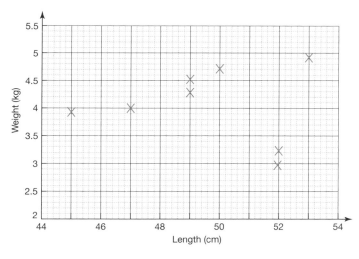

a Draw a ring round any outliers.

b Draw a line of best fit on the diagram.

c What type of correlation is displayed in the diagram?

d Use your line of best fit to estimate the weight of a baby measuring 51 cm.

e Explain why you wouldn't use your line of best fit to predict the length of a baby weighing 5.5 kg.

2 In an experiment seedlings are grown different distances from a light source. The distance from the light source and height of the seedling after 2 weeks are recorded.

Distance from light source (cm)	Height (cm)
0	4.2
25	3.9
50	3.8
75	2.5
100	3.0
125	0
150	2.2

a Plot a scatter graph to show the data.

b Draw a line of best fit on the data.

c Draw a ring round any values you consider to be outliers and explain why these might have occurred.

d Copy and complete the sentence:

The further the seedling is from the light source, the...

3 An economics researcher says, 'There is positive correlation between the number of betting shops in an area and the crime rate.'

Do you think this correlation implies causation? (That is, does the number of betting shops cause the crime rate to be higher?) Explain your answer.

You may **not** use a calculator for these questions.

1 Henry carries out a survey about people's favourite TV programme. He asks 10 of his friends.

Give two reasons why his sample is biased.

2 India carried out a survey of 20 students to find out what their favourite pizza toppings were.

The table gives information about her results.

Topping	Frequency
Margherita	11
Vegetarian	2
Hawaiian	6
Pepperoni	1

a Which topping is the mode?

b In India's survey, what percentage of people like pepperoni best?

c India is going to draw a pie chart to display her information. What angle should the sector for pepperoni be?

3 The pie chart below shows the colour of electric cars sold by a garage one year.

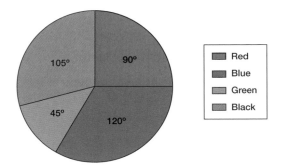

a What fraction of the cars sold were red?

60 green cars were sold.

b How many cars does the whole pie chart represent?

c How many black cars were sold?

4 The graph gives information about the percentage of people who do their grocery shopping online.

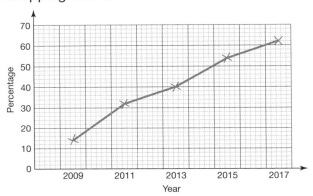

a Describe the trend in the percentage of people who do their grocery shopping online.

b Use your graph to predict the percentage of people who will shop online in 2019.

c Is your predication reliable? Explain your answer.

5 The mean, median, mode and range of a set of four numbers is calculated.

Mean = 6

Mode = 7

Median = 7

Range = 6

Which of the following statements are true?

a All of the numbers are positive.

b At least one of the numbers is 7.

c Two of the numbers are 7.

d Three of the numbers are 7.

6 The table shows some information about the pets owned by students.

	Cat	Dog	Fish	Rabbit
Girls	3	5	1	0
Boys	8	2	2	4

a Draw a suitable diagram or chart for this information.

b What fraction of students have a cat?

7 The table shows information about the age of customers using a library.

Age	Frequency
0–19	17
20–39	2
40–59	32
60–79	23
80–99	9

a What is the median class?

b Alin calculates the range of ages by working out 99 − 0 = 99
The range may not be 99.
Explain why.

8 A shoe shop records the size of shoes sold one morning. The information is shown in the bar chart.

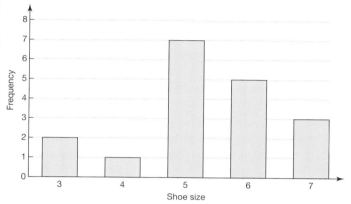

a How many size 5 shoes were sold?

b What is the mode of the sizes of shoe sold?

c Calculate the mean size of shoe sold.

d Which measure of central tendency is most useful for the manager to know? Give a reason for your answer.

9 This pictogram shows the A level subjects studied by a group of science students:

Subject	Number of students
Physics	◯ ◯
Chemistry	◯ ◯ ◖
Biology	◯ ◯ ◯

◯ = 20 students

a What fraction of the students studied chemistry?

b How many more students studied biology than physics?

c What is the modal subject studied?

10 The ages of 9 people and the number of minutes they spend using a mobile phone each day is recorded.

Age	18	23	45	17	32	65	56	36	40
Minutes	62	54	15	60	25	13	67	28	16

a Draw a scatter graph to show the data.

b What type of correlation is shown?

c Put a ring around any outliers.

d Use your line of best fit to predict how many minutes a 30 year old would spend on the phone.

e Explain why you would be unlikely to use this graph to predict how long a 70 year old would spend on the phone.

f Describe in words the relationship between the time spent on the phone and the age of a customer.

You **may** use a calculator for these questions.

11 The table shows the weight of 10 patients at a doctors' surgery.

Weight, w (kg)	Frequency
$60 \leq w < 70$	3
$70 \leq w < 80$	5
$80 \leq w < 90$	2

a Calculate an estimate of the mean weight.

b Explain why this is only an estimate.

12 A sample of 5000 people are surveyed about their annual income.

Annual income	Percentage of population
$0 \leq x < £25\,000$	35%
$£25\,000 \leq x < £40\,000$	55%
$£40\,000 \leq x < £80\,000$	10%

Draw a pie chart to represent the data.

13 An athlete records the length of time it takes her to run 100 m during 10 different training sessions. She works out her mean time is 10.3 seconds.

The following day she runs 100 m in 9.5 seconds. What is her new mean time?

14 Catherine rolls a dice 5 times. Her mean score is 3.8. The range of scores is 4. The modal score is 3.

What numbers did she roll?

Answers

For full worked solutions, visit:
www.scholastic.co.uk/gcse

Number

Factors, multiples and primes

1 a 5   **b** 1, 12   **c** 1, 5, 45   **d** 1, 5

2 HCF = 10, LCM = 1050

3 $2 \times 3^2 \times 5$

4 a 10   **b** 840

5 12 and 18

Ordering integers and decimals

1 a false   **c** true   **e** false
   b true   **d** false

2 $-0.3, -1.5, -2.5, -4.2, -7.2$

3 0.049, 0.124, 0.412, 0.442, 1.002

4 a <   **b** <   **c** >

Calculating with negative numbers

Stretch it! negative, yes

1 a −11   **c** −6   **e** 0
   b 99   **d** 18   **f** 25

2 −8 and 9

3 32°C

Multiplication and division

Stretch it! 148 419

1 a 2115   **b** 56 364

2 a 47   **c** 126 remainder 4 or $126\frac{4}{17}$
   b 516

3 a 33 boxes   **b** 1 pencil

4 £91.25

5 £288

6 $307\frac{2}{3}$

7 28 805

8 37 boxes

9 He has not placed a zero in the ones column before multiplying through by 5.

Calculating with decimals

Stretch it! 18.2

1 a 2.33   **c** 0.035   **e** 1.563
   b 24.391   **d** 6.099

2 £4.64

3 Erica: £54.92; Freya: £27.46

Rounding and estimation

Stretch it! a 1.0   **b** 1.00   **c** 1.000 − they are all 1

Stretch it! 55.25 m² − an overestimate.

1 a 0.35   **c** 32.6
   b 10   **d** 33 100

2 a $150 \le x < 250$ **c** $3.15 \le x < 3.25$
   b $5.5 \le x < 6.5$ **d** $5.055 \le x < 5.065$

3 $\frac{30}{0.5 \times 6} = 10$

4 b is false since $18 \times 1 = 18$ so 18×0.9 cannot be 1.62

c is false since if you divide by a number smaller than 1 the answer will be larger.

5 Tarik should choose One tariff.

Converting between fractions, decimals and percentages

Stretch it! $0.\dot{1}, 0.\dot{2}, 0.\dot{3}, \ldots 0.\dot{4}, 0.\dot{5}$

1 a $\frac{32}{100} = \frac{8}{25}$ **c** $\frac{33}{100}$
   b $1\frac{24}{100} = 1\frac{6}{25}$ **d** $\frac{95}{100} = \frac{19}{20}$

2 a $0.41\dot{6}$ **c** 0.49   **e** $0.\dot{4}2857\dot{1}$
   b 0.375   **d** 0.185

3 a 91%   **c** 80%
   b 30%   **d** 60%

4 37.5%

5 30%, 0.35, $\frac{2}{5}$

6 $\frac{15}{20} = \frac{75}{100} = 75\%$ – Amy

Rudi was highest

Ordering fractions, decimals and percentages

1 $\frac{7}{12}, \frac{3}{8}, \frac{1}{3}$

2 $-2.2, -\frac{1}{10}, 1\%, 0.1, 15\%, \frac{1}{5}, 7$ (so the middle is 0.1)

3 Yes, if the numerator of a fraction is $\frac{1}{2}$ the denominator the fraction is equivalent to $\frac{1}{2}$. If the numerator is smaller than this the fraction must be smaller than $\frac{1}{2}$.

Calculating with fractions

Stretch it! No, you could add the whole number parts then the fraction parts, giving:

$1 + 2 = 3$

$\frac{3}{5} + \frac{1}{4} = \frac{17}{20}$

$\qquad = 3\frac{17}{20}$

1 a $1\frac{5}{8}$ **c** $\frac{10}{21}$ **e** $\frac{2}{25}$
   b $\frac{6}{17}$ **d** $8\frac{3}{20}$

2 a 12   **b** £35   **c** 808 mm

3 20

4 35

Percentages

1 a 1.8 cm   **b** £0.30   **c** 4 ml

2 a 33   **b** 540   **c** £101.92

3 a 480   **b** 133   **c** £14.58

4 3052

5 £14 300

Order of operations

1 a 7   **b** −1.9   **c** −13

2 30

3 $(8 - 3 + 5) \times 4$

Exact solutions

1 **a** π **b** 36π **c** $2\frac{1}{2}\pi$

2 **a** 7π **b** $\frac{5}{8}\pi$

3 Area $= \frac{6}{28} = \frac{3}{14}$ cm². Perimeter $= 2\frac{1}{14}$ cm

4 **a** 18π cm   **b** 144π cm²

5 $\frac{1}{2}\pi$ cm

Indices and roots

Stretch it!

a Multiplying by 2

b Multiplying by $\frac{3}{2}$

Stretch it! 3.7 to 1 decimal place

1 **a** $\frac{1}{3}$ **b** $2\frac{1}{2}$ **c** $1\frac{1}{9}$

2 $1^3, \sqrt[3]{8}, \sqrt[3]{27}, 3^2$

3 **a** -8 **c** 81

     b 1 **d** 1

4 **a** $\frac{1}{4}$ **c** 1

     b $\frac{1}{49}$ **d** $\frac{1}{3}$

5 5^4

6 $\frac{1}{12}$ $\sqrt[3]{7}$ 3.7 $\sqrt{20}$

7 $\sqrt{196} = \sqrt{2^2 \times 7^2} = 14$

Standard form

1 **a** $45\,000\,000$ **b** 0.091

2 **a** 6.45×10^8 **b** 7.9×10^{-8}

3 $345\,800$

4 3.1×10^{-2} 3.09×10 $3 + (2.1 \times 10^2)$ 3.2×10^2

5 3×10^8

6 $0.022 = 2.2 \times 10^{-2}$ m $= 2.2$ cm

Listing strategies

Stretch it!

red + small, red + medium, red + large,
green + small, green + medium, green + large,
blue + small, blue + medium, blue + large.

1 111

112, 121, 211, 113, 131, 311

222
221, 212, 122, 223, 232, 322

333
331 313 133 332 323 233

123 132 213 231 312 321

2 444 446 449
464 466 469
494 496 499

3 Small A, Small B, Small C, Small D
Medium A, Medium B, Medium C, Medium D
Large A, Large B, Large C, Large D.

Review it!

1 7 and 6 (or 11 and 2, where both are prime and 2 is also a factor of 12)

2 $620 = 2 \times 2 \times 5 \times 31 = 2^2 \times 5 \times 31$

3 $18 = 2 \times 3 \times 3$
$36 = 2 \times 2 \times 3 \times 3$
$40 = 2 \times 2 \times 2 \times 5$
HCF $= 2$

4 $-11.5, -8.3, -3.5, -3.2, 1.4$

5 **a** £51.73
     b £18.33
     c $51.73 - 32.99 = 18.74$ **or** $51.73 - 18.74 = 32.99$
        $18.33 \times 3 = 54.99$

6 **a** 3.22 **b** 4023

7 27

8 $31\frac{4}{11}$

9 **a** 0.375 **b** 70%

10 **a** $\frac{7}{10}$ **b** $\frac{4}{5}$

11 All of them.

12 **a** $\frac{26}{35}$ **b** $1\frac{1}{2}$ **c** $1\frac{1}{2}$

13 $0.25 - 0.07$

14 $\frac{3}{4}$

15 $\frac{9}{200}$

16 £2.15

17 **a** 9   **b** 5

18 5 and 6

19 **a** 3.4×10^9 **b** 3.04×10^{-7}

20 $37.55 \le x < 37.65$

21 **a** 51   **b** 12, 15, 21, 51, 25, 52

22 **a** $200 \times 9 \times 10 = 18\,000 = £180.00$
     b Underestimate since all numbers were rounded down.

23 240

24 35%

25 no since 2 is a prime number and odd + odd + even = even

26 **a** 1.08×10^{11} **b** 3×10^2

27 £279.20

28 **a** 3.1 **b** 3.05

29 **a** $325\,000$ **b** $320\,000$

30 729

31 £16.62

32 3420

33 **a** 2010 and 2011   **b** $1.1 \times 102.3 = 112.53$

Algebra

Understanding expressions, equations, formulae and identities

1. **a** $3a + 6 = 10$ **c** $3(a + 2)$
   b $C = \pi D$ **d** $3ab + 2ab = 5ab$

2. James is correct.

   $4x - 2 = 2x$ can be solved to find the value of x so it is an equation.

   Or, the two sides of $4x - 2 = 2x$ are not equal for all values of x so it cannot be an identity.

Simplifying expressions

Stretch it! $12t \times t \times t$, $2t \times 6t \times t$, $2t \times 3t \times 2t$, $3t \times 4t \times t$

1. **a** p^3 **c** $12ab$ **e** $-8g^2$
   b $28bc$ **d** $20x^2$ **f** $6pqr$

2. **a** $5x$ **c** 6 **e** $4x$
   b $-7w$ **d** $4n$ **f** $9a$

Collecting like terms

1. **a** $5f$ **h** $-3a - b - 3$
   b $7b$ **i** $2x^2$
   c $5mn$ **j** t^3
   d $3a + 1$ **k** $2a + b^2$
   e $4d - 2e$ **l** $7\sqrt{x}$
   f $5x + 3y - 2$ **m** $3\sqrt{x}$
   g $7a + 5b$ **n** $7\sqrt{x}$

Using indices

1. **a** x^9 **c** $6m^8$ **e** u^3
   b p^5 **d** $15m^6n^4$ **f** t

2. **a** x^2 **d** $2x^3$ **g** $\frac{x}{3}$
   b y^4 **e** $\frac{1}{m^2}$
   c p **f** $\frac{x^4}{3}$

3. **a** x^6 **c** p^{10} **e** $\frac{1}{x^6}$
   b y^{16} **d** $16m^{10}$ **f** n^8

4. x^4

Expanding brackets

Stretch it! **a** $a\sqrt{3} + a^2$ **b** $b\sqrt{5} - b^2$ **c** $c + d$

Stretch it!

1. $(x + 2)(x + 4) = x^2 + 6x + 8$

2. **a** $2x^2 + 8x + 6$ **c** $6x^2 + 7x - 3$
   b $3x^2 + 10x - 8$ **d** $6x^2 - xy - y^2$

1. **a** $3a + 6$ **e** $4x + 4y + 8$
   b $4b - 16$ **f** $-2y - 4$
   c $10c + 25$ **g** $x^2 - 2x$
   d $6 - 2e$ **h** $2a^2 + 10a$

2. **a** $3a - 5$
   b $6x + 4$

3. **a** $8x + 26$ **b** $y - 3$ **c** $2m + 31$

4. **a** $x^2 + 5x + 6$ **c** $a^2 - 4a - 21$
   b $y^2 + y - 12$ **d** $m^2 - 7m + 6$

5. **a** $x^2 + 2x + 1$ **c** $m^2 - 4m + 4$
   b $x^2 - 2x + 1$ **d** $y^2 + 6y + 9$

Factorising

Stretch it! $(x + 1)$

1. **a** $3(a + 3)$ **c** $7(1 + 2c)$
   b $5(b - 2)$ **d** $d(d - 2)$

2. **a** $4(2a + 5)$ **c** $9(2 + c)$
   b $4(b - 3)$ **d** $d(2d - 3)$

3. **a** $2(2x - 3y)$ **e** $n(2 - 9n)$
   b $m(a + b)$ **f** $5x(1 + 2y)$
   c $4a(3a + 2)$ **g** $4p(q - 3)$
   d $x(4x + 3y)$ **h** $4y(x^2 - 2)$

4. $a = 2$, $b = 3$

5. **a** $(x + 1)(x + 7)$ **e** $(x - 3)(x - 3)$
   b $(x - 1)(x + 5)$ **f** $(x + 3)(x + 4)$
   c $(x + 2)(x - 4)$ **g** $(x - 2)(x + 5)$
   d $(x - 2)(x - 3)$ **h** $(x + 4)(x - 5)$

6. **a** $(x + 4)(x - 4)$ **c** $(x + 9)(x - 9)$
   b $(x + 6)(x - 6)$ **d** $(y + 10)(y - 10)$

Substituting into expressions

1. 11

2. **a** 10 **c** -22 **e** 28
   b -24 **d** 20 **f** -1

3. False.
   When $a = 3$: $3a^2 = 3 \times 3^2 = 3 \times 9 = 27$.

4. **a** -20 **c** -8
   b 2 **d** 26

5. -3.5

Writing expressions

1. **a** $4 - q$ **c** xy (or yx)
   b $n + m$ (or $m + n$) **d** p^2

2. $x + y$

3. $\frac{y}{8}$

4. $100n + 75b$

5. $9a + 2$

6. $4a + 10$

Solving linear equations

1. **a** 7 **c** 16
   b 17 **d** -2

2. **a** 5 **c** -4 **e** 11
   b 5 **d** 33 **f** 19

3. **a** 2 **c** $\frac{5}{2}$ (Or 2.5, or $2\frac{1}{2}$)
   b 3 **d** -2

4. Hannah has not subtracted 4 from **both** sides.

5. **a** 3 **d** -4
   b 3 **e** $-\frac{4}{8} = -\frac{1}{2}$ (Or -0.5)
   c 3

6. **a** 3 **d** 3
   b 3 **e** -6
   c $\frac{5}{2}$ (Or 2.5 or $2\frac{1}{2}$)

7. **a** -4 **c** 1
   b 2 **d** $\frac{9}{6} = \frac{3}{2}$ (Or 1.5 or $1\frac{1}{2}$)

Writing linear equations

1 a $8s + 12$ **b** 9 cm

2 a 50   **b** 115°

3 30 years old

4 4

5 8 cm

6 $a = 18$, $b = 28$

Linear inequalities

1 a $x = 3, 4, 5$ **c** $x = 0, 1, 2, 3$

  b $x = 2, 3, 4, 5$ **d** $x = -3, -2, -1, 0, 1$

2 a $x < 3$ **c** $-1 \le x \le 5$

  b $x \ge -2$

3 a

  b

  c

  d

4 a $x > 3$

  b $x \le 5$

  c $x \le \dfrac{10}{4} = \dfrac{5}{2}$

  $\left(\text{Or } x \le 2.5 \text{ or } x \le 2\tfrac{1}{2}\right)$

  d $x < -5$

  e $x > -4$

  f $x \ge -1$

5 a $x = -2, -1, 0, 1, 2$

  b $x = -4, -3, -2, -1, 0, 1, 2, 3, 4, 5, 6$

  c $x = 0, 1, 2, 3$

  d $n = 3, 4, 5$

6 a $3 \le x$ (or $x \ge 3$)   **c** $\tfrac{1}{2} \ge x$ $\left(\text{or } x \le \tfrac{1}{2}\right)$

  b $-1 > x$ (or $x < -1$)   **d** $2 > x \ge -3$

Formulae

1 Wage earned = £305

2 113°F

3 -90

4 $C = 25d + 50$

5 a $k = \dfrac{8}{5}m$ **b** 320 km

6 a $4a + 6$ **b** 30 cm

7 -65

8 a $\dfrac{v - u}{t} = a$ **d** $\dfrac{v^2 - u^2}{2a} = s$

  b $\dfrac{3V}{A} = h$ **e** $g = \dfrac{2s}{T^2}$

  c $\dfrac{y + 9}{3} = x$

  $\left(\text{Or } x = \dfrac{y}{3} + 3\right)$

Linear sequences

1 a 27, 33   **b** add 6

2 a 1st term = 2, 2nd term = 6,
    3rd term = 10, 4th term = 14

  b 78

3 a

Pattern number	1	2	3	4	5
Number of sticks	7	10	13	16	19

  b 25

4 a 5, 7, 9   **b** 203

  c 35th term has a value of 73

5 a 3, 10   **b** 8th term

6 a $4n - 1$

  b $4n - 1 = 100$

  $4n = 101$

  $n = 25.25$

  100 cannot be in the sequence because 25.25 is not an integer

Non-linear sequences

1 b   **3** 0.0005   **5** 6, 9, 14, 21

2 27, 38   **4** 42

Show that...

Stretch it! n^2 will be even $\times$ even = even,
so $n^2 - 1$ will be odd.
m^2 will be odd $\times$ odd = odd,
so $m^2 - 1$ will be even.

1 a LHS $= 4(x - 3) + 2(x + 5) = 4x - 12 + 2x + 10$
        $= 6x - 2$

    RHS $= 3(2x - 1) + 1 = 6x - 3 + 1 = 6x - 2$

    LHS = RHS

    So $4(x - 3) + 2(x + 5) \equiv 3(2x - 1) + 1$

  b LHS $= (x + 2)(x - 2) = x^2 - 2x + 2x - 4 = x^2 - 4$

    LHS = RHS

    So $(x + 2)(x - 2) \equiv x^2 - 4$

Number machines

1 a $y = 3x - 3$ **b** 27   **c** $x = \dfrac{y + 3}{3}$

  d If $y = x$,

    $3x - 3 = x$

    $2x = 3$

    $x = \dfrac{3}{2}$

Coordinates and midpoints

Stretch it!

Difference in x coordinates of A and $B = 2 - (-1) = 3$

So x coordinate of $C = -1 - 3 = -4$

Difference in y coordinates of A and $B = 4 - 3 = 1$

So y coordinate of $C = 3 - 1 = 2$

1 a $(2, 2)$

b and **c**

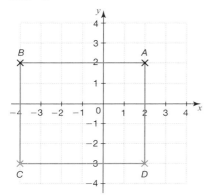

d $(-1, -0.5)$

Straight-line graphs

Stretch it! $x = 0.5$

1

2

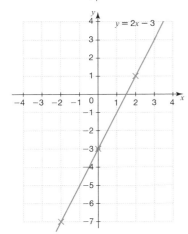

3 $y = x + 1$

4 $y = 2x - 4$

5 $y = -\frac{1}{2}x + 4$

6 a

b

7 $4x + 4y = 8$

$4y = 8 - 4x$

$y = 2 - x$

$2x = 6 - 2y$

$2y = 6 - 2x$

$y = 3 - x$

Both lines have a gradient of -1 so they are parallel.

8 Gradient, $m = -2$

y-intercept is $(0, 4)$

9 a $y = 3.4$ **c** $x \approx 2.5$

   b $x = 0.5$ **d** $x \approx -0.5$

Solving simultaneous equations

1 $x = 10.5, y = 5.5$

2 a $x = 1, y = 2$ **d** $x = 3, y = 2$

   b $x = 3, y = -2$ **e** $x = 4, y = -1$

   c $x = 6, y = -4$ **f** $x = 2, y = 1$

3 7 and 14

4 A burger costs 95 p. A cola costs £1.10.

5 a $x = 2, y = 8$ **b** $x = 8, y = 1$

Quadratic graphs

Stretch it! $x \approx 3.6$ and $x \approx -0.6$

1 a

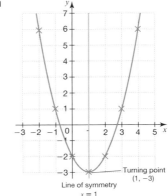

b $x = 1$ **c** $(1, -3)$

2 a

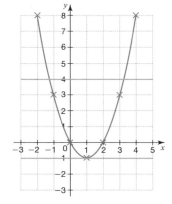

b $x = 0$ and $x = 2$

3 a

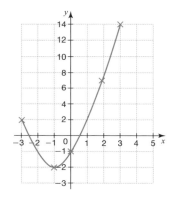

b $x \approx -2.4$ and $x \approx 0.4$

Solving quadratic equations

Stretch it!

$x = 4$ or $x = -4$

$x = 5$ or $x = -5$

1 a $x = 0$ or $x = 4$ **d** $x = -1$ or $x = -9$

  b $x = 0$ or $x = -7$ **e** $x = 3$ or $x = -4$

  c $x = -4$ or $x = 4$ **f** $x = -2$ or $x = 8$

2 a $x = -7$ or $x = 7$

  b $x = 0$ or $x = 3$

  c $x = -1$ or $x = -6$

Cubic and reciprocal graphs

1 a

x	−3	−2	−1	0	1	2	3
y	−28	−9	−2	−1	0	7	26

b

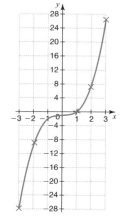

2 a

x	−4	−3	−2	−1	$-\frac{1}{2}$	$-\frac{1}{4}$	$\frac{1}{4}$	$\frac{1}{2}$	1	2	3	4
y	$\frac{1}{4}$	$\frac{1}{3}$	$\frac{1}{2}$	1	2	4	−4	−2	−1	$-\frac{1}{2}$	$-\frac{1}{3}$	$-\frac{1}{4}$

b

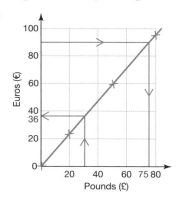

Drawing and interpreting real-life graphs

1 a

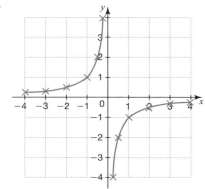

b i €36   **ii** £75

c From the graph: £30 = €36

So £90 = €36 × 3 = €108

Ring is cheaper in France.

2 a £10   **b** 13p

3 a

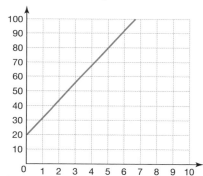

b £20

c Let $d = 6$ days (more than 5 days)

First shop: $C = 12 \times 6 + 20 = 92$

Second shop: $C = 10 \times 6 + 30 = 90$

To hire the sander for more than 5 days use the second shop as it is cheaper.

4 a 30 minutes

  b 55 km

  c Speed before break = 30 km/hr

Speed after break = 16.7 km/hr

d

5 a 12.8 m

b 3 seconds

c 0.4 seconds and 2.5 seconds

d The ball is thrown from a height of 3 m above the ground.

6 a

b $0.\dot{3}$ m/s²

7 a 35 cm

b The person stayed in the bath.

c The person got out of the bath.

d Running water into the bath was quicker. The slope of the line between O and A (filling the bath) is steeper than the slope of the line between E and F (emptying the bath).

Review it!

1 a (3, 5)

b, c and **e**

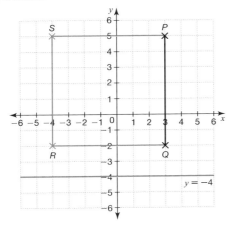

d (−0.5, 1.5)

2 a −2

b C

c Millie is correct.

When $x = 4$, $3x^2 = 3 \times 4^2 = 3 \times 16 = 48$

3 a $4a - 4$ **c** m^5

b $4(2x + 3)$ **d** x^5

4 a

x	−1	0	1	2	3
y	−1	1	3	5	7

b

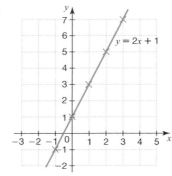

c 2

5 $x = 3$

6 a $x \leq 2$

b

c $x = 1, 2, 3$

d $x \leq \frac{3}{2}$

7 a $6x$ or $x + 65$

b $6x = x + 65$

$5x = 65$

$x = 13$

Luke is 13 years old.

8 a 27, 33

b No. This is a sequence of odd numbers and 44 is even.

c 47

9 $x^2 + 7x + 12$

10 $a - b = 12$

11 $x = 4, y = 2$

12 $a = 2, b = 3$

13 a $12m$ **b** $12p^2$ **c** $6x$

14 a $w = 11$ **b** 21   **c** $5a + 8b$

15 a (−5, 4)

b Length = 8 units, width = 6 units

16 a

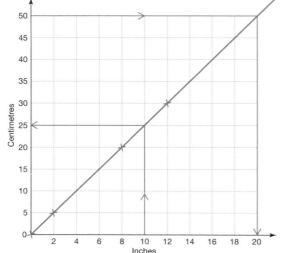

  b i 25 cm   **ii** 20 inches

  c £3.00

17 a $T = 12.50x + 10$ **b** 5 days

18 $4(P - R) = Q$

19 a $m(m + 8)$ **b** $(x + 3)(x + 4)$

20 a nth term $= 3n - 1$

  b 50th term $= 149$

  c $2n - 3 = 112$

    $2n = 115$

    $n = 57.5$

    112 cannot be a term in the sequence because 57.5 is not an integer.

21 a $-2x + 23$ **b** $20a^5b^3$

22 $x = 4.5$

23 Input $= -2$

24 $a = 6$

25 C

Ratio, proportion and rates of change

Units of measure

1 a 3000 m   **c** 13 000 cm²   **e** 7200 seconds

   b 75 mins   **d** 3.52 litres   **f** 14 kg

2 4.175 kg or 4175 g

3 2.27 kg

Ratio

Stretch it! $\frac{31}{56}$

1 a 1:4   **b** 1:3:4   **c** 4:5

2 7:1

3 Allow 1 part cement for 1.5 parts sand.

4 a 7:1   **b** 100 tickets

5 a 3:2   **b** 120

6 0.6 kg

7 12 cm and 15 cm

8 $s = 20t$

9 40 g

Scale diagrams and maps

Stretch it! $50 \div x$

1 A, B, F

2 a 36 km   **b** 1.25 cm

3 120 m

4 a 1 km   **b** 250°

Fractions, percentages and proportion

1 $\frac{1}{175}$

2 $\frac{11}{24}$

3 a $\frac{3}{4}$ **b** 25%

4 10%

5 School A: 125:145 = 25:29
School B: 100:120 = 5:6
No since the ratios are not equivalent.

6 4.125 g

Direct proportion

Stretch it! Since for two values to be in direct proportion when one is zero the other must be zero.

1 A and E

2 a i 30 meringues   **b** 7 eggs
    ii 100 meringues

3 2 hours 30 mins

4 A, D

Inverse proportion

1 D

2 22.5 mins

3 $1\frac{2}{3}$ of a day.

4 a 2

  b The age of the chicken and the number of eggs it lays are in inverse proportion. This means that as the age of the chicken increases so the number of eggs it lays decreases.

Working with percentages

Stretch it! £128
Stretch it! 2% ($1.02^5 \times 100 = £110.41$)

1   **a**   £51.50   **b**   992   **c**   12.48
2   12.5%   **3**   20.9°C   **4**   25920
5   £150   **6**   88.9%

Compound units

Stretch it! $\frac{100}{x}$ mph

1   164 units   **3**   0.24 g/cm³
2   40 minutes   **4**   6 N/m²
5   10.8 km/hour
6   475 gallons per hour (to the nearest whole number)
7   Bolt: 100 m in 9.58 seconds = 10.4 m/s
    Cheetah: 120 km/hr = 120 000 m/hour = 120 000
    ÷ 60 m/min = 2000 m/min = 2000 ÷ 60 m/sec
    = 33.3 m/s
    The cheetah is faster.

Review it!

1   **a**   3200 m   **b**   540 seconds   **c**   400 ml
2   4.6 km
3   150 minutes
4   3520 cm² or 0.352 m²
5   30 000 cm²
6   $\frac{5}{12}$
7   13 : 9
8   5 minutes
9   2300 kg/m³
10   25%
11   $\frac{12}{25}$ or 48%
12   $\frac{23}{90}$
13   25 hours = 1 day and 1 hour
14   **a**   30 more   **b** $C = 4J$
15   28%
16   £864
17   6 cm
18   £1591.81
19   8000 people
20   3 hours 9 minutes
21   £15.75
22   20%
23   £3675
24   18 : 13
25   1750 men
26   No — for two things to be in direct proportion when one is zero the other must be zero. The graph does not go through the origin so this is not the case.
27   Neither, since the time taken to cook increases as the weight increases it is not in indirect proportion. It is not in direct proportion since a graph to illustrate the relationship would not go through the origin.
28   11 seconds
29   She is incorrect since the ratio of females to males must be the same for them to have equivalent proportions.
30   225 g

Geometry and measures

Measuring and drawing angles

1   **a**   43°   **b**   Acute
2   **a**

    b

    c

3   **a**

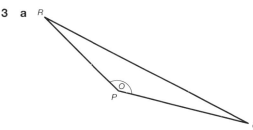

    b   18°

Using the properties of angles

1   54°
2   **a**   70°
      Base angles of an isosceles triangle are equal.
    b   110°
3   18°
4   **a**   54°
      Angles on a straight line add up to 180°.
    b   83°
5   **a**   96°
    b $y = 96°$
      Use the fact that corresponding angles are equal, then vertically opposite angles are equal.
      Or, use the fact that alternate angles are equal, then use angles on a straight line add up to 180°.
6   **a**   58°   **b**   64°
    c   58° (Alternate angles are equal **or** opposite angles of a parallelogram are equal)
7   Angle $BAD = 62°$ (Opposite angles of a parallelogram are equal)
    Angle $ADE = 62°$ (Alternate angles are equal)
    $x = 180 - 62 - 62$ (Base angles of an isosceles triangle are equal)
    $x = 56°$

8 Angle $ACB = 36°$ (Base angles of an isosceles triangle are equal)

Angle $ABC = 180 - 36 - 36$ (Angles in a triangle add up to $180°$)

Angle $ABC = 108°$

$x = 108°$ (Alternate angles are equal)

Using the properties of polygons

Stretch it!

1 The angle sum of a triangle is $180°$.

Sum of interior angles of a hexagon $= 4 \times 180° = 720°$.

2

Polygon	Number of sides (n)	Number of triangles formed	Sum of interior angles
Triangle	3	1	180°
Quadrilateral	4	2	360°
Pentagon	5	3	540°
Hexagon	6	4	720°
Heptagon	7	5	900°
Octagon	8	6	1080°
Decagon	10	8	1440°

3 $n - 2$

4 $180 \times (n - 2)$

Stretch it!

Angles around a point add up to $360°$.

No, $360°$ is not divisible by $108°$ (interior angle of a regular pentagon).

1 $36°$

2 a 24   **b** $3960°$

3 Angle $CBE = 72°$

Using bearings

1 a $315°$

b

2 $344°$

3 Kirsty is correct.

The bearing is $314°$ ($360° - 46°$) as it must be measured clockwise from North.

Properties of 2D shapes

Stretch it!

1 a

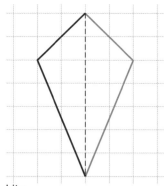

b kite

2 a 8 possible lines of symmetry:

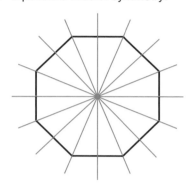

b 8

3 a 2   **c** 1, no

b rhombus   **d** square, rhombus

Congruent shapes

1 Any accurate copy of shape A.

2 a 120°   **b** 12 cm

3 a SSS   **b** ASA

Constructions

Stretch It!

5 cm

60°

1

2

3

4

5

Drawing circles and parts of circles

1 **a** chord   **d** arc
    b tangent   **e** radius
    c diameter   **f** segment

2

3

Loci

1

2

3 **a** and **b**

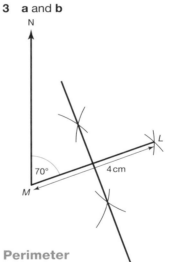

Route taken by boat

Perimeter

1 28.8 cm
2 42 cm
3 $k = 4, b = 8$
4 $200 + 30\pi$ m
5 £1.80

Area

Stretch it! Area of a semicircle $= \frac{\pi r^2}{2}$,
Area of quarter circle $= \frac{\pi r^2}{4}$

1 **a** 9.0 cm²   **c** 28.0 cm²   **e** 63.6 cm²
    b 4.5 cm²   **d** 5.0 cm²
2 9 cm²
3 $\frac{1}{4}$
4 18 cm²
5 454.1 cm²

Sectors

1 Area = 39.3 cm²
   Perimeter = 25.7 cm
2 Area = 12π cm²
3 £119.92

3D shapes

Stretch it!

3D shape	Faces	Edges	Vertices
Cube	6	12	8
Cuboid	6	12	8
Square-based pyramid	5	8	5
Tetrahedron	4	6	4
Triangular prism	5	9	6
Hexagonal prism	8	18	12

Stretch it!

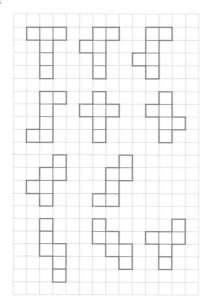

1 a 6 possible rectangular faces:

b

c Kelli has not counted the hidden edges.

2

3 a

   b

4
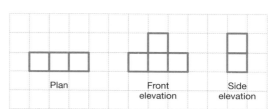

| Plan | Front elevation | Side elevation |

5 a

   b

Volume

1 382 cm³
2 1.96 m³
3 k = 264
4 6.3 cm

Surface area

1 150 cm²
2 36π cm²
3 14 cm²
4 76.3%

Using Pythagoras' theorem

1 10.2 cm
2 6 m
3 54 cm²
4 If the triangle is right-angled, $PQ^2 = PR^2 + RQ^2$
   $PQ^2 = 13^2 = 169$
   $PR^2 + RQ^2 = 8^2 + 5^2 = 69$
   $PQ^2 \neq PR^2 + RQ^2$
   Claudia is not correct.

5 10.63

6 $AB = 6.7$ (1 d.p.)

7 £910

Trigonometry

Stretch it! Any lengths for the opposite and hypotenuse in the ratio 1:2.

1 **a** 0.4 **c** 1.0 **e** 48.6
    b 0.6 **d** 26.6 **f** 54.7

2 4.6 cm

3 40.6°

4 $\text{Sin } 15° = \frac{\text{opposite}}{10}$
    Opposite = 2.6 m

Exact trigonometric values

1 **a** 0.5 **c** 0 **e** $\sqrt{3}$
    b 0 **d** $\frac{1}{\sqrt{2}}$

2 $AC = 4$ cm
    $BC = 4\sqrt{2}$ cm

3 30° and 60°

4 $\sin 30° = \frac{1}{2}$ therefore, $ABC = 30°$

5 0.5, $\frac{3}{4}$, cos 30°, tan 45°

Transformations

1

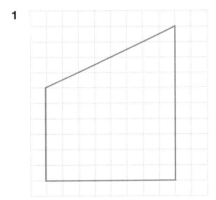

2 Translation by vector $\begin{pmatrix} -4 \\ -2 \end{pmatrix}$

3 **a** and **b**

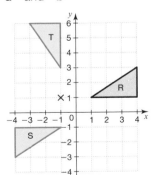

4 Reflection in the y-axis

5 Enlargement by scale factor $\frac{1}{2}$, centre (3, 3)

6 **a** and **b**

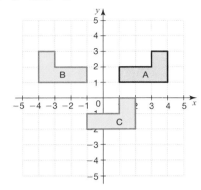

c Rotation of 90° clockwise about (0, 0)

Similar shapes

Stretch it! Perimeter of triangle $ABC = 14$ cm; perimeter of triangle $DEF = 28$ cm. The perimeter of a shape enlarged by scale factor 2 will also be enlarged by scale factor 2. When a shape is enlarged by any scale factor, the perimeter of the shape is enlarged by the same scale factor.

1 **a** 30° **b** 16 cm **c** 2 cm

2 **a** 80° **b** 13.2 cm **c** 4 cm

Vectors

1 **a**

c

b

d

2 **a**

b

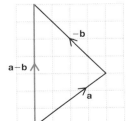

c $\begin{pmatrix} -8 \\ -6 \end{pmatrix}$

3 a $\begin{pmatrix} -3 \\ 9 \end{pmatrix}$ **c** $\begin{pmatrix} 7 \\ 3 \end{pmatrix}$

  b $\begin{pmatrix} -8 \\ 12 \end{pmatrix}$ **d** $\begin{pmatrix} -9 \\ -9 \end{pmatrix}$

4 a 4**a**

  b $-3\mathbf{b}$

  c $4\mathbf{a} - 3\mathbf{b}$

  d $-3\mathbf{b} - 4\mathbf{a}$

Review it!

1 a

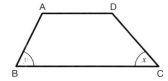

  b 3.8 cm   **c** 50°

2 a 5

  b 6

3

4 A shape that matches the shape given in the question. It could be on its side or upside down.

5 72 cm²

6 24 cm

7

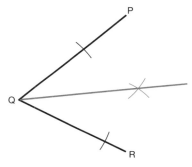

8 Rotation of 180° about (1, 0)

9 $x = 130°$

10 $\frac{9}{20}$ or 45%

11 If triangle *ABC* is right-angled, $c^2 = a^2 + b^2$

  $c^2 = 8^2 = 64$

  $a^2 + b^2 = 6^2 + 4^2 = 36 + 16 = 52$

  $c^2 \neq a^2 + b^2$ so triangle *ABC* is not right-angled.

12

13 a $\frac{1}{\sqrt{2}} = \frac{\sqrt{2}}{2}$ **b** $AB = 3$ cm

14 a $\begin{pmatrix} 8 \\ 1 \end{pmatrix}$ **b** $\begin{pmatrix} -6 \\ 13 \end{pmatrix}$

15

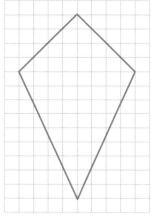

16 Equilateral triangle

17 a 35°

     Base angles of an isosceles triangle are equal.

  b 110°

  c Triangle *XYZ* is isosceles so angle *XYZ* = angle *YXZ*. Angle $c = 55°$

18 12.6 cm

19 135°

20

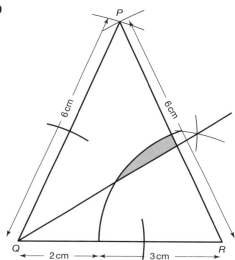

21 29.06 cm

22 32.3 cm

23 7.7 cm²

24 4 times

25 Length = 13.2 cm

26 $DE = 7.2$ cm

27 a 300 cm²

  b 300 cm³

28 53.1°

29 Translation by vector $\begin{pmatrix} -7 \\ -6 \end{pmatrix}$

30 $5\mathbf{a} + 2\mathbf{b}$

Probability

Basic probability

Stretch it! No – each time the probability of getting an even number is $\frac{1}{2}$. You would expect to get even numbers approximately 50 times but cannot guarantee it.

1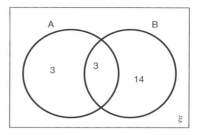

2 a $\frac{3}{25}$ **b** $\frac{13}{25}$

3 a

4 b

5 0.43

6 0.4

Two-way tables and sample space diagrams

1 a 12

b

	Chicken	Beef	Vegetarian
Fruit	12	6	4
Cake	5	3	8
Total	17	9	12

2 a

		Dice 1					
		1	2	3	4	5	6
Dice 2	1	2	3	4	5	6	7
	2	3	4	5	6	7	8
	3	4	5	6	7	8	9
	4	5	6	7	8	9	10
	5	6	7	8	9	10	11
	6	7	8	9	10	11	12

b i $\frac{1}{18}$ **ii** $\frac{1}{12}$ **iii** 0

3 1, 1, 3, 3

Sets and Venn diagrams

Stretch it! None

1 a

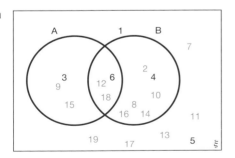

b Multiples of 6

2 a C ∩ T is the set of students who travel by car **and** train.

C′ ∩ B is the set of students who do **not** travel by car **and** travel by train.

b i $P(C) = \frac{19}{50}$

    ii $P(B \cup T) = \frac{3}{5}$ **iii** $P(B' \cap T) = \frac{7}{25}$

3

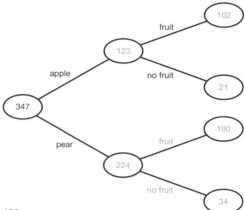

Frequency trees and tree diagrams

1 a

b $\frac{190}{347}$

2 a

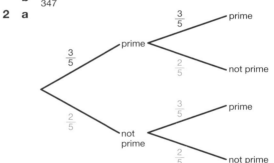

b $\frac{9}{25}$ **c** $\frac{21}{25}$

3 a

b $\frac{5}{7}$

Expected outcomes and experimental probability

Stretch it! The dice has not been rolled enough times – carry out further tests.

1 135

2 20 red sweets

3 50 primes

4 a Charlie – he has carried out the most tests.

    b 6

1 30

2

3 0.7

4 B, C

5 a $\frac{3}{5}$

  b 5

6

	Pizza	Pasta	Risotto	Total
Cake	12	6	1	19
Ice cream	10	11	10	31
Total	22	17	11	50

7 0.7

8 a No, he has not tested his dice enough times.

  b 18

9 0.55

10 a

		Dice					
		1	2	3	4	5	6
Coin	Head	2	4	6	8	10	12
	Tail	3	4	5	6	7	8

  b i $\frac{1}{6}$ **ii** $\frac{1}{6}$

11 a i 6   **ii** 1   **iii** 5

  b $\frac{4}{8} = \frac{1}{2}$

12 a

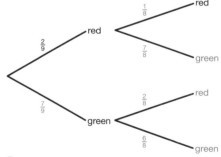

  b $\frac{7}{12}$

13 a 10

  b $\frac{1}{3}$ is only a theoretical probability and therefore will not necessarily be accurate in real life.

14 0.74 or $\frac{37}{50}$

15 £50

16 a Milo, he has surveyed a larger sample.

  b 263

Statistics

Data and sampling

Stretch it! A random sample could be taken; you could allocate a number to each student and randomly generate the numbers to survey. Any method is acceptable as long as each person in the school has an equally likely chance of being chosen. Alternatively a stratified sample could be taken.

1 Primary source: Recording the data by measuring it yourself.

  Secondary source: Any sensible source, e.g. the Met Office, local paper, etc.

2 Qualitative data.

3 It is cheaper and quicker than surveying the whole population.

4 a The people working for an animal charity are more likely to be opposed to wearing real fur. Every member of the population does not have an equal chance of being chosen.

  b Surveying people in the street, a random telephone survey. Any sensible method that ensures that each member of the population has an equal chance of being chosen.

5 a 12 000

  b The sample is relatively small. The sample is not a random sample as it is taken on one day in a year.

Frequency tables

1

Number of people on the bus	Frequency
1–9	4
10–19	12
20–29	3
30–39	1

2 a

Number of courgettes	Frequency
0	1
1	0
2	1
3	1
4	9
5	3
6	0

  b 56

3 a There are gaps between the groups.

  The groups do not have equal width.

  b Although one of the data values may fall in the $30 \le h < 40$ group, this doesn't mean that they trained for 40 hours. They could have trained for any length of time between 30 and 40 hours.

Bar charts and pictograms

1 a 20   **b** 25%

2 a 4   **b** 39.2%

  c Proportion of boys who play two sports $= \frac{6}{18} = \frac{1}{3}$

  Proportion of boys who play three sports $= \frac{3}{12} = \frac{1}{4}$

  $\frac{1}{3} > \frac{1}{4}$ so the proportion who played two sports is larger.

3

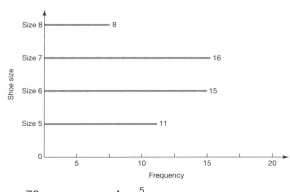

4 a 70   **b** $\frac{5}{21}$

Pie charts

Stretch it! Round appropriately – but check the angles sum to 360°.

1 Spanish = 168°
French = 108°
German = 84°

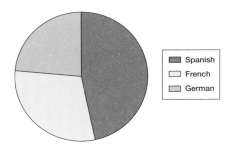

2 a 1   **b** 97.5% or $\frac{39}{40}$

3 a 28

  b The bar chart, since the frequency is easy to read from the bar chart.

Measures of central tendency: mode

1 The other three must be 12.2.

2 $1 < t \le 2$

3 Max is correct. The modal number of pets is the group with the highest frequency, therefore 2 pets is the mode.

Measures of central tendency: median

1 5.4

2 $2 < b \le 4$

3 If there are 5 integers in the list then the middle value is the 3rd integer.

Measures of central tendency: mean

Stretch it! a Mode **b** Mean/Median **c** Mean/Median

1 a Mean = 4.6

  b You are using the midpoint of the groups as an estimate of the actual value for each group.

2 8.5

3 No – they could be any pair of numbers which sum to 10.

Range

1 8.8

2 a Girls = 3

  b Boys = 2

3 Athlete A

4 35% or 75%

Comparing data using measures of central tendency and range

1 a i Mean = 83.5 minutes

    ii Median = 30.5 minutes

  b The extreme values affects the mean but not the median.

2 All the data is used to find the mean.

3 Either as long as suitably justified:

Car A – although the mean time is higher, it is more consistent in performance since the range is smaller.

Car B – the acceleration is quicker on average.

4 a and **b** The mode or median since the mean will not be a whole number and therefore not meaningful.

Time series graphs

1

2 a 67°C

  b Approx. 27°C

  c No, since it is extrapolation (beyond the limits of the data).

3 a 17 000

  b i April   **ii** August

  c The number of tourists peaks in April and again in December. The low seasons are February/March and July/August/September.

Scatter graphs

1 a and **b**

c Positive

d This will vary according to the line of best fit: approximately 4.8 kg

e This is beyond the limits of the data and therefore extrapolation.

2 **a, b** and **c**

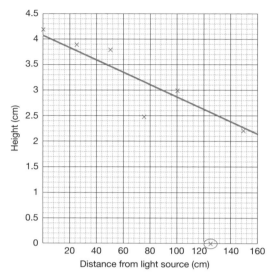

c The seeds failed to germinate or the seedling died.

d The further the seedling is from the light source the shorter its height.

3 No, although the two things correlate one does not cause another. There may be many reasons why the crime rate is high in the area, perhaps there is poverty and inequality causing social tension.

Review it!

1 The sample is too small and he only asked his friends, therefore not representative of the population of TV viewers.

2 **a** Margherita   **b** 5%   **c** 18°

3 **a** $\frac{1}{4}$

b 480 cars

c $\left(\frac{105}{360}\right) \times 480 = 140$ cars

4 **a** The number of people doing their grocery shopping online is increasing.

b Any sensible answer: approximately 75%

c No − it is outside the limits of the data therefore extrapolation.

5 **a** True

b True

c True

d False

6 Comparative bar chart or compound bar chart:

a

or

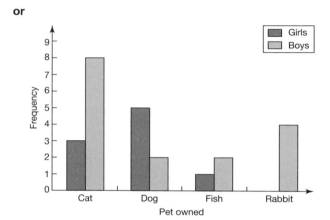

b $\frac{11}{25}$

7 **a** 40−59

b The youngest person is between 0 and 19, the youngest may be any age in this range and the oldest is between 80 and 99, therefore, any age in this range.

8 **a** 7

b Size 5

c Mean = 5.$\dot{3}$

d Mode − the mean is not an actual shoe size.

9 **a** $\frac{1}{3}$

b 20

c Biology

10 a and **c**

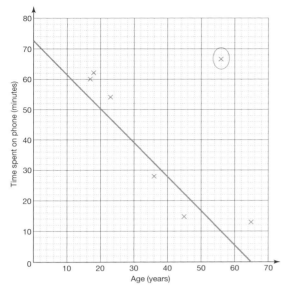

b Negative

d Approximately 40 minutes: depends on line of best fit.

e This is outside the limits of the data and therefore extrapolation.

f As the age of the customer increases the time spent on the phone decreases.

11 a 74

b The midpoint of the class is used as the age of each of the patients rather than the actual age.

12 Annual income for surveyed population

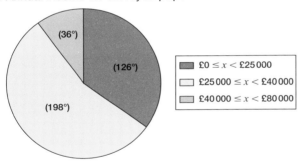

13 Mean = 10.2 (1 decimal place)

14 2, 3, 3, 5, 6

Number
Factors, multiples and primes

(1) Work out Highest Common Factor (HCF) of 18 ◄——
and 24. (2 marks, ★★)

For HCF: List all the factors for each number. Now look for the highest factor in **both** lists.

....

(2) Which of these are prime numbers?

Circle your answers. (2 marks, ★★)

15 16 17 18   19 20 21 22 23 24 25

Give your answer in index form, like this: $60 = \square^{\square} \times \square \times \square$

(3) Write 60 as a product of its prime factors. (2 marks, ★★★) ◄——

....

(4) In a piece of music there are two drummers.

Drummer 1 hits her drum after 6 seconds.

She then hits her drum every 6 seconds after that.

Drummer 2 hits his drum after 8 seconds.

Drummer 2 hits his drum every 8 seconds after that.

How many times do the drummers hit their drums
at the same time in the first minute?

List all the times Drummer 1 will hit her drum in the first minute: 6, 12, … seconds. Then list all the times Drummer 2 will hit his drum in the first minute.

Show working to justify your answer. (3 marks, ★★★★★) ◄——

....
....
....
....
....

STRETCHIT!

Can you write a factor tree for the number 216?

Ordering integers and decimals

(1) Write these numbers in order of size. Start with the smallest number. (1 mark, ★)

-8   -12   2   0   -1

....

(2) Write these decimals in descending order. (1 mark, ★★)

0.32   0.203   0.23   0.3

....

SNAPIT! **Ascending and descending**

Ascending order means from smallest → largest.
Descending order means from largest → smallest.

(3) Fill in the gaps with > or <. (★★★)

a -4 ☐ 0.4 (1 mark)

c -0.404 ☐ -0.44 (1 mark)

b 4.200 ☐ 4.3 (1 mark)

d 0.33 ☐ 0.4 (1 mark)

[Total: 4 marks]

SNAPIT! **Less than or greater than**

< means less than; > means greater than.

Calculating with negative numbers

(1) Work out (★)

   a $-7 + -3$ (1 mark)

   ....

   b $-7 - -3$ (1 mark)

   ....

   c $8 + -5 - -2$ (1 mark)

   ....

   d $-4 - -6 + -1$ (1 mark)

   ....

[Total: 4 marks]

SNAP IT! Adding or subtracting negative numbers

Adding a negative number is the same as subtracting. This means you can rewrite the calculation as a subtraction.

Subtracting a negative number is the same as adding a positive number. This means you can rewrite the calculation as an addition.

(2) Work out (★)

   a -9×2 (1 mark)

   ....

   b $-12 \div -3$ (1 mark)

   ....

   c $-4 \times -2 \times 5$ (1 mark)

   ....

   d $-24 \div 3 \times 2$ (1 mark)

   ....

[Total: 4 marks]

SNAP IT! Multiplying or dividing negative numbers

When multiplying and dividing negative numbers
- if the signs are the same, then the answer is positive
- if the signs are different, then the answer is negative.

STRETCH IT!

What would happen if you multiplied three negative numbers like $-2 \times -3 \times -4$?

(3) Which of these numbers is 4 **less** than -7, **multiplied** by -2?

Circle your answer. (2 marks, ★★)

   -6 11   6   22

(4) There are five questions in a quiz. People score 3 marks for each correct answer and -2 marks for each incorrect answer.

Sally scores -5. How many correct and incorrect answers did she get? (3 marks, ★★★★)

....

Multiplication and division

(1) Work out (★)

a 357 × 6 (1 mark)

c 552 ÷ 6 (1 mark)

....

....

b 261 × 43 (1 mark)

d 676 ÷ 13 (1 mark)

....

....

[Total: 4 marks]

> **NAILIT!**
>
> Read the question carefully. Decide if you need to multiply or divide.

(2) 300 textbooks are packed into boxes. 24 books fit in each box. (★★)

a How many boxes are filled? (2 marks)

....

b How many books are left over? (2 marks)

....

[Total: 4 marks]

(3) Jamie takes out a car loan for £12 500.

His first payment is £440.

After that, he makes 36 equal repayments.

How much is each repayment? (2 marks, ★★★)

....

(4) Clara's working contract states:

Working hours: 26 hours per week.

Holiday: 6 weeks per year (including bank holidays). ◄—— There are approximately 52 weeks in a year.

How many hours does Clara work each year? (2 marks, ★★★★)

....

Calculating with decimals

Estimate the answer by working out 9×8.
Now you know approximately what your answer should be.

(1) Work out 9.2×8.3 (2 marks, ★★) ◄

....

(2) Aisha buys a dress for £22.50 and some shoes for £19.99.

She pays with a £50 note.

How much change should she get? (2 marks, ★★)

....

WORKIT!

Sam and Jamie visit the funfair. Sam spends three times as much as Jamie. The total amount they spend is £35.60.

How much do they each spend?

Diagrams can help you solve maths problems. This bar shows how much Jamie spends.

This bar shows how much Sam spends (because Sam spends three times as much as Jamie).

This bar shows the total amount they spend.

Jamie spends: £35.60 ÷ 4

£8.90

Sam spends: £8.90 × 3

£26.70

(3) Work out $229.74 \div 6$ (2 marks, ★★★)

....

(4) Kirsty and Flo are raising money for charity. Kirsty raises five times as much as Flo. The total amount they raise is £172.50. How much do they each raise? (3 marks, ★★★)

....

Rounding and estimation

(1) Round 0.7983 to (★★)

a 3 significant figures (1 mark)

....

b 2 decimal places. (1 mark)

....

[Total: 2 marks]

(2) Estimate an answer to this calculation. (2 marks, ★★★)

$$\frac{9.74 \times 4.02}{7.88}$$ ⟵ Round each number to the nearest whole number. Then work out the calculation.

....

(3) A concert ticket costs £39.50.

489 concert tickets are sold.

The concert organisers have costs of £12 500. (★★★)

a Estimate the amount of profit the organisers make. (3 marks)

....

b Is your answer to part **a** an underestimate or an overestimate? (1 mark)

Give a reason for your answer.

....

....

[Total: 4 marks]

SNAPIT! **Profit and estimation**

Profit is the income (amount made) minus any costs.
Underestimate means to estimate something as less than it really is.
Overestimate means to estimate something as more than it really is.

Converting between fractions, decimals and percentages

1 Fill in the gaps. (★★)

   a 0.071 as a fraction is 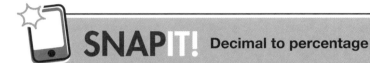 (1 mark)

   b 63% as a decimal is _____ (1 mark)

   c 0.4 as a percentage is _____% (1 mark)

> ## SNAP**IT!** Decimal to percentage
>
> To convert a decimal to a percentage multiply by 100.
> **Beware!**
> 0.1 as a percentage is 0.1 × 100 = 10% (**not** 1%!)
> 0.01 as a percentage is 0.01 × 100 = 1%

   d 32% as a fraction in its simplest form is _____ (1 mark) ◀── Cancel the fraction.

[Total: 4 marks]

2 Work out $\frac{5}{16}$ as (★★★)

   a a decimal (1 mark)

_____

   b a percentage. (1 mark)

_____

[Total: 2 marks]

3 Which of these numbers is the largest? Circle your answer. (2 marks, ★★★)

$\frac{5}{8}$ 0.65   60%

Ordering fractions, decimals and percentages

(1) Fill in the gaps with > or <. (★)

a $\frac{1}{2}$ ☐ 0.6 (1 mark)

b $\frac{3}{4}$ ☐ 0.7 (1 mark)

c $-\frac{3}{10}$ ☐ 0.2 (1 mark)

[Total: 3 marks]

(2) Write in ascending order (★★)

a $\frac{5}{12}$ $\frac{7}{15}$ $\frac{9}{20}$ (1 mark)

....

b 45% $\frac{1}{25}$ 0.4   (1 mark)

....

[Total: 2 marks]

 SNAPIT! **Ordering fractions**

> To order fractions, convert them all to a common denominator.
>
> To order fractions, decimals and percentages, convert them all to decimals or percentages.

(3) A bag in shop A is reduced by $\frac{1}{3}$

The same bag in shop B is reduced by 30%

The same bag in shop C is reduced by $\frac{2}{5}$

Write the shops in order, starting with the shop that offers the bag at the cheapest price.

Show working to justify your answer. (2 marks, ★★★)

 NAILIT!

Sometimes you don't need to convert. You can use logical thinking instead.

For example, is $\frac{5}{9}$ bigger or smaller than $\frac{1}{2}$ (i.e. 0.5)? How does it compare to the other fraction, decimal and percentage in the list?

....

(4) Write in descending order (2 marks, ★★★★)

0.38 $\frac{5}{9}$ 38.5% $\frac{3}{10}$

....

Calculating with fractions

(1) Work out $\frac{1}{5} + \frac{4}{9}$ (2 marks, ★)

....

(2) Work out $2\frac{3}{4} - 2\frac{2}{3}$ (2 marks, ★★) ◄——— Write mixed numbers as improper fractions.

....

(3) Work out $1\frac{5}{6} \times \frac{2}{7}$ (2 marks, ★★) ◄——— Cancel down answers if you can.

....

(4) Jo has a 6 m length of silver chain. She uses $\frac{3}{5}$ m to make one necklace.

How many necklaces can she make from the chain? (2 marks, ★★★★)

Use easier numbers to help you decide what to do. If Jo used 2 m to make one necklace, what calculation would you do?

....

Percentages

① 40% of students in a class play a musical instrument. There are 25 students in the class. How many play a musical instrument? (2 marks, ★)

> Work out an easy percentage first, like 10%. 40% = 10% × ▢

② A scarf is £16. A shop offers a 15% discount in its sale.

How much is the scarf now? (2 marks, ★★)

③ A theme park predicts a rise in visitors of 14% next year. This year it had 12 450 visitors. How many does it predict for next year? (2 marks, ★★★)

④ A decorator charges £40 per hour. He also charges VAT at 20% on all his bills. How much is his bill for 3 days' work, if he works 7 hours per day? (3 marks, ★★★★)

> VAT is Value Added Tax. It is a tax that is paid on the price of some goods and services. Currently, it is 20% in the UK.

Order of operations

(1) Work out $3^2 - 6 \div (2 + 1)$ (1 mark, ★)

....

SNAP IT! BIDMAS

Do operations in calculations in this order:

Brackets   **I**ndices   **D**ivision or **M**ultiplication   **A**ddition or **S**ubtraction

(2) Work out $2^3 + 3 \times \sqrt{25}$ (1 mark, ★★)

....

(3) Work out the value of $(1.7 - 0.12)^2 + \sqrt[3]{4.096}$ (2 marks, ★★★★)

....

SNAP IT! Roots on a calculator

For $\sqrt[3]{}$ on your calculator: enter the number to cube root, then press the $\sqrt[y]{x}$ key and enter 3. Finally, press =.

Exact solutions

(1) Work out the area of the triangle. Give an exact solution. (1 mark, ★)

0.76 cm
0.35 cm

SNAPIT! **Exact solution**

Giving an exact solution means **not** rounding.
You may give an exact solution as a decimal (**not** rounded), fraction, surd, or in terms of π.

Remember, a surd is a square root that can't be simplified.

....

(2) Work out the area of the square. Give an exact solution. (2 marks, ★★)

$1\frac{1}{3}$ m

$1\frac{1}{3}$ m

....

(3) Work out the area of the rectangle. Give an exact solution. (2 marks, ★★★★)

√6 cm

√2 cm

An exact solution will mean you leave surds in your answer, so it will be in this form ▮ √ ▮

....

(4) Show that the area of this shape is 3π cm². (3 marks, ★★★★★)

2 cm

DOIT!

'Show that' means show the working that leads to a particular answer.

1 Write down: area of a circle = πr^2

2 Write down: the fraction of the circle shown = ▢

3 Write down: the area of the circle shown = ▢
(the fraction × the formula for the area of a circle)

4 Write down: the radius, r = ▢

Show that when you substitute the radius (from 4) into your formula (from 3) and then simplify, you get the exact solution 3π cm².

Indices and roots

1. Write these in index form. (★★)

   a $7 \times 7 \times 7 \times 7$ (1 mark)

   ....

   b $\dfrac{1}{5 \times 5 \times 5}$ (1 mark)

   ....

   [Total: 2 marks]

2. Write down the value of (★★★)

   a 2^4 (1 mark)

   ....

   b 10^{-2} (2 marks)

   ....

   [Total: 3 marks]

3. Write the following in order of size, starting with the smallest.

   2^3 3^{-2} $\sqrt{25}$ $\sqrt[3]{27}$

   Show working to justify your answer. (3 marks, ★★★★)

   ....

4. Simplify $\dfrac{9^5}{9^3 \times 9^2}$ (2 marks, ★★★★)

   ....

Standard form

(1) Write 2.75×10^3 as an ordinary number. (1 mark, ★)

....

(2) The average distance from the Sun to the Earth is 150 000 000 km.

Write this number in standard form. (2 marks, ★★★)

NAILIT!

For standard form, write the digit as a number between 1 and 10, and then multiply by a power of 10.

....

(3) Write 0.00642 in standard form. (2 marks, ★★★)

....

(4) A snail travels 1.4×10^{-5} km in 1 minute. How far does it travel in 20 minutes?

Write your answer in standard form. (2 marks, ★★★★★)

....

Listing strategies

(1) Write down all the ways these cards can be arranged. (2 marks, ★★)

| 2 | 5 | 9 |

....

....

....

> When listing, work logically. For this question
> - list all the ways the cards can be arranged when the first card is 2
> - now list all the ways the cards can be arranged when the first card is 5
> - now list all the ways the cards can be arranged when the first card is 9.

(2) Pip spins these 3-sided and 4-sided spinners, then adds the scores. (★★★)

a Complete the sample space diagram to show all possible outcomes. (2 marks)

		4-sided spinner			
		0	**1**	**2**	**3**
3-sided spinner	**1**	1	2		
	2	2			
	3				

b How many different ways can Pip score 3? (1 mark)

....

[Total: 3 marks]

(3) Terence is wrapping a birthday present. He has red, blue and green wrapping paper and five different types of gift tags.

How many different combinations of wrapping paper and gift tags can Terence use? (2 marks, ★★★)

....

(4) A children's menu offers

Starter – soup (s) or bread (b)

Main – pasta (p) or fish fingers (f)

Dessert – jelly (j) or ice cream (i)

List all the different options. Use the initials given in brackets. (3 marks, ★★★★)

Algebra
Understanding expressions, equations, formulae and identities

① Write down whether each of these is an equation, an expression or an identity. (★)

a $\frac{6u}{3} \equiv 2u$ (1 mark) ....

b $p + 7 = 12$ (1 mark) ....

c $8x - y + 5xy$ (1 mark) ....

[Total: 3 marks]

② Write down whether each of these is an equation, an expression or a formula. Give a reason for each answer. (★★★)

a $2a - 3 = 25$ (1 mark)

....

b $F = ma$ (1 mark)

....

c $6m + a$ (1 mark)

....

d $k = \frac{1}{2}mv^2$ (1 mark)

....

SNAP IT! Expressions, equations, identities

- An expression can contain letter terms and/or numbers. It does **not** have an equals (=) sign.
- An equation contains letter terms and numbers. It does have an equals (=) sign.
- You can solve an equation to find the value of the letter.
- An identity is true for all values of the letters. It is sometimes written with a sign like this ≡.

NAIL IT!

When asked to 'Give a reason for your answer', write your answer, then write 'because...'. Finish the sentence with a reason.

[Total: 6 marks]

SNAP IT! Formulae

A formula can contain letter terms and/or numbers. It does have an equals (=) sign.
The letters represent values.
These values can vary. For example, the formula for the area of a triangle $= \frac{1}{2} \times$ base $\times$ height or $A = \frac{1}{2} bh$.
The values of bh vary according to the lengths of the sides of the triangle.

③ Here are five cards. The first two show numbers. The next shows a letter. There is also the operation 'add' and an 'equals sign'. (★★★)

| 2 | 10 | x | + | = |

a Use four cards to write an expression. (1 mark)

....

b Use four cards to write an equation. (1 mark)

....

[Total: 2 marks]

Simplifying expressions

(1) Simplify $9x - 4x + 3x$ (1 mark, ★)

> $9x$, $4x$ and $3x$ are like terms, because they all contain a number and the letter x.
> You can simplify an expression by adding and subtracting like terms.

....

(2) Simplify (★★★)

a $6a \times 8a$ (1 mark)

> $a \times a = a^2$
> $a \times a \times a = a^3$

....

b $2p \times 3p \times 5p$ (1 mark)

....

[Total: 2 marks]

(3) Simplify $35yz \div 7z$ (1 mark, ★★★)

....

(4) Simplify $\frac{32uv}{4v}$ (1 mark, ★★★)

> Cancel the numbers; then cancel any letters.
> What are you left with?

....

STRETCHIT!

You can use exactly the same logic for questions that look harder but are not.

Imagine you have four piles of 1p coins:

101p + 102p + 103p + 194p

Collect like terms to find out how many pence (p) you have. Convert that into pounds (£).

Collecting like terms

(1) Simplify (★★)

   a $7m + 6n - 4m - 2n$ (1 mark) ◄─── Rearrange the expression so that all the like terms are together $7m - 4m$ and $6n - 2n$.

....

   b $9q - 5r - 12q + 3r$ (1 mark)

....

[Total: 2 marks]

NAILIT!

Remember that each term has a sign in front of it. Don't forget the sign.

(2) Simplify (★★)

   a $11a + 5b - 10a + 8b$ (1 mark)

....

   b $6c - 4d - 7c + 5d$ (1 mark)

....

[Total: 2 marks]

NAILIT!

You never write 1 in front of a letter in algebra. Instead, you just write the letter. For example, write a **not** $1a$.

(3) Simplify (★★★)

   a $9p^3 + p - 4p^3$ (1 mark) ◄─── To add or subtract terms, they must be **exactly** the same. p^3 and p are **not** like terms.

....

   b $12 - 5x^2 + 3x - 2x^2$ (2 marks) ◄─── Usually terms are written with the highest power first. So $\_\_x^2 + \_\_x + \_\_$

....

   c $m + m + m + n \times n$ (1 mark) ◄─── Look carefully at the operations.

[Total: 4 marks]

....

(4) Simplify $3\sqrt{5} - f - 8\sqrt{5} + 2f$ (2 marks, ★★★★)

....

SNAPIT! **Adding and subtracting surds**

You can add or subtract surds where the square root number is the same.

For example,

$3\sqrt{2} - 7\sqrt{2} = -4\sqrt{2}$

$\sqrt{2} + 8\sqrt{2} = 9\sqrt{2}$

Using indices

(1) Simplify (★★)

   a $p^3 \times p$ (1 mark)

....

   b $4y^2 \times 3y^3$ (2 marks) ◄

> Rearrange the expression so that it is easier to see what you need to do.
> $4 \times y^2 \times 3 \times y^3$
> $= 4 \times 3 \times y^2 \times y^3$

....

   c $2a^4b \times 5ab^2$ (2 marks)

....

[Total: 5 marks]

(2) Simplify (★★★)

   a $q^{-2} \times q^{-4}$ (1 mark)

....

   b $(u^{-3})^2$ (1 mark)

....

   c $x^{-1} \times x$ (1 mark)

....

[Total: 3 marks]

(3) Simplify (★★★)

   a $b^4 \div b^3$ (1 mark)

....

   b $\dfrac{f^5}{f^2}$ (1 mark)

....

   c $\dfrac{xy^3}{x^2y}$ (2 marks)

....

[Total: 4 marks]

(4) Complete $\left(\boxed{}\, m^3\right)^{\square} = 8\,m^9$ (2 marks, ★★★★)

....

Expanding brackets

(1) Expand (★★)

   a $4(m + 3)$ (1 mark)

....

> To expand a bracket, multiply **every** term inside the bracket by the term outside the bracket.

   b $2(p - 1)$ (1 mark)

....

NAILIT!

Remember that each term has a sign in front of it. Don't forget the sign, when multiplying.

   c $10(3x - 5)$ (1 mark)

....

[Total: 3 marks]

(2) Expand and simplify (★★★)

   a $3(m + 2) + 5(m + 1)$ (2 marks)

....

   b $6(x - 1) - 2(x - 4)$ (2 marks) ← Once you have expanded the bracket, don't forget to simplify.

....

[Total: 4 marks]

(3) Expand and simplify (★★★★)

   a $(v + 3)(v + 7)$ (2 marks)

....

NAILIT!

To expand two brackets, multiply **every** term inside one bracket by **every** term inside the other bracket.

Where there are two terms in each bracket, you should get an answer with four terms.

Simplify to get an answer with three terms.

   b $(b + 2)(b - 4)$ (2 marks)

....

   c $(x - 4)(x - 6)$ (2 marks)

....

[Total: 6 marks]

(4) Expand and simplify (★★★★★)

   a $(q + 1)^2$ (2 marks) ←

> Rearrange the expression so that there are two brackets. For example, $(q + 1)^2 = (q + 1)(q + 1)$

....

   b $(c - 3)^2$ (2 marks)

....

   c $(4m - n)(3m + n)$ (2 marks)

....

[Total: 6 marks]

Factorising

(1) Fully factorise (★★)

  a $4x + 8$ (1 mark)

....

  b $3d - 15$ (1 mark)

....

  c $8y - 12$ (1 mark)

....

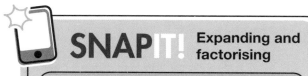
SNAP IT!   **Expanding and factorising**

Expanding removes the brackets.

$3(x - 2) = 3x - 6$

Factorising inserts the brackets.

[Total: 3 marks]

(2) Factorise (★★★)

  a $q^2 + q$ (1 mark)

....

  c $10z^2 + 15z$ (2 marks)

....

  b $a^2 + 6a$ (1 mark)

....

> **NAIL IT!**
>
> Fully factorise means make sure you use the highest factor outside the bracket and not a smaller one.

[Total: 4 marks]

(3) Factorise (★★★★)

  a $x^2 + 7x + 12$ (2 marks)

....

  b $x^2 + 6x - 16$ (2 marks)

....

  c $a^2 - 10a + 24$ (2 marks)

....

> Write $(x\ \ )(x\ \ )$
> Write all the factor pairs that make 12, including any negative ones. For example, 1, 12; –1, –12; 2, 6…
> Which factor pair adds to the number in the x term, 7?

[Total: 6 marks]

(4) Factorise (★★★★)

  a $y^2 - 4$ (2 marks) ◄ The y term is 0.

....

  b $x^2 - 9$ (2 marks)

....

  c $p^2 - 100$ (2 marks)

....

[Total: 6 marks]

Substituting into expressions

(1) Work out the value of $4x + 5y$ when $x = 3$ and $y = -2$ (2 marks, ★)

....

(2) $s = ut + \dfrac{1}{2}at^2$ ◄——— Use BIDMAS for the order of operations (see page 254).

Work out the value of s when $u = 12$, $a = 10$ and $t = 2$ (2 marks, ★★★)

....

(3) $c = 7$ and $d = -5$. Work out the values of f when (★★★★)

a $f = 3c - 2(c - d)$ (1 mark)

....

b $f = -c(d^2 - 3c)$ (2 marks)

....

c $f^2 = 7c - 3d$ (2 marks) ◄——— There are two possible values of f for part c. Make sure you write both.

....

[Total: 5 marks]

(4) Work out the value of $\dfrac{p(q - 3)}{r}$

when $p = 6$, $q = 7$ and $r = -8$ (2 marks, ★★★)

....

Writing expressions

(1) **a** Ali thinks of a number, n. He adds 3 to get a new number.

Write an expression for Ali's new number. (1 mark, ★★)

> Pretend Ali's original number is 7. What calculation would you do to work out his new number?
> Now do the same for the letter n. This is the expression for Ali's new number.

....

b Kamal thinks of a number, n. He doubles it and subtracts 9 to get a new number.

Write an expression for Kamal's new number. (2 marks, ★★)

....

[Total: 3 marks]

(2) Eve sells hats for £x and scarves for £y. Write an expression for the money she makes when she sells (★★★)

a 1 hat and 1 scarf (1 mark)

....

c 12 hats and 11 scarves. (2 marks)

....

b 5 hats (1 mark)

....

[Total: 4 marks]

(3) A parallelogram has side lengths $9p$ and $5p + 2$.

Write an expression for its perimeter.

Give your answer in its simplest form. (2 marks, ★★★★)

> Often it can help to draw a diagram. Draw the parallelogram and put all the information on it.

....

(4) A rectangular flower bed has side lengths s and $5s + 1$.

Write an expression for its area. (1 mark, ★★★★)

> Use a bracket in your expression.

....

Solving linear equations

(1) Solve these equations. (★★)

a $x + 5 = 12$ (1 mark)

....

> To get x on its own on one side of the equals sign, you need to subtract 5. You must do the same to both sides.

NAILIT!

Solving equations is about getting the letter on one side of the equals sign and a number on the other side.

b $x - 3 = 10$ (1 mark)

c $4x = 20$ (1 mark)

d $\frac{x}{3} = 6$ (1 mark)

....

....

....

[Total: 4 marks]

(2) Solve these equations. (★★)

a $2x + 3 = 15$ (2 marks)

....

> First of all, get $4x$ on its own on one side of the equals sign. Then get x on its own.

c $\frac{x}{5} + 3 = 8$ (2 marks)

....

> First of all, add $2x$ to both sides of the equation, so that your x term is positive.

b $3x - 5 = 16$ (2 marks)

d $7 - 2x = 1$ (2 marks)

....

....

[Total: 8 marks]

(3) Solve these equations. (★★★)

a $3(x + 9) = 30$ (2 marks)

....

> First of all, expand to remove the brackets.

c $2(10 - 3m) = 8$ (2 marks)

....

b $5(p - 2) = 10$ (2 marks)

d $4(6 - 2q) + 10 = 2$ (2 marks)

....

....

[Total: 8 marks]

(4) Solve these equations. (★★★★)

a $4x - 6 = x + 9$

(2 marks)

> First of all, deal with the smallest x term. Subtract this from both sides.

c $4(2x + 3) = 11x + 3$

(3 marks)

> First of all, expand to remove the brackets.

....

....

b $2y + 5 = 4y - 3$ (2 marks)

d $3(n + 4) = 2(2n + 3)$ (3 marks)

....

....

[Total: 10 marks]

Writing linear equations

① Here is a triangle. What is the value of x? (2 marks, ★★)

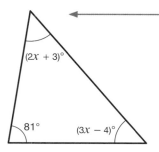

(2x + 3)°

81°   (3x − 4)°

> What do the angles in a triangle add to? Write an equation and solve it to find x.

....

② Sophie is half the age of Jamie. The sum of their ages is 18.

Work out Jamie's age. (2 marks, ★★★)

> Start by choosing a letter to represent the unknown: Sophie's age = a
> Now read one sentence at a time, and try writing an expression or equation for the new information.
> For example, Jamie's age = ▇ (Make sure half of your answer is Sophie's age, a.)

....

③ The length of a rectangle is 3 cm longer than the width.

The perimeter of the rectangle is 46 cm.

Work out the area of the rectangle. (3 marks, ★★★)

DOIT!

> Draw a diagram of the rectangle and put all the information on it.

....

④ Here is a parallelogram.

(5x − 20)°
(7x + 5y)°

(3x + 10)°

DOIT!

> Write down everything you know about angles in a parallelogram.

Work out the values of x and y. (3 marks, ★★★★)

....

Linear inequalities

(1) x is an integer such that $-2 < x \leq 5$ (★★)

   a Write down all the integers that satisfy
      the inequality. (1 mark)

   ....

   b Show the inequality $-2 < x \leq 5$ on a number line. (2 marks)

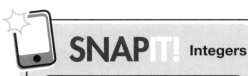

SNAP IT! Integers

An integer is any whole
number. It can be positive,
negative, or zero.

[Total: 3 marks]

(2) Solve each inequality. Show each answer on a number line. (★★★)

   a $4x > 20$ (2 marks)

   b $3x - 8 \leq 13$ (2 marks)

   c $2(x - 3) < 10$ (2 marks)

[Total: 6 marks]

(3) Solve (★★★★)

   a $2 \leq 3x + 5 < 11$ (2 marks)

   Write and solve two inequalities:

   $2 \leq 3x + 5$ and $3x + 5 < 11$

   Give your answer in the form $\square \leq x < \square$

   ....

   b $-4 < 5x + 6 \leq 6$ (2 marks)

   ....

[Total: 4 marks]

(4) Fran says she is thinking of a positive integer, n.
   She says if you sum her integer, and 3 more   Write an inequality and solve it.
   than her integer, then the answer is less than 15.
   List all possible integer values for n. (2 marks, ★★★★★)

   ....

Formulae

(1) The length of time, in minutes, to roast a chicken is calculated using this formula

$t = 40w + 20$

where t is the time in minutes and w is the weight of the chicken in kg. (★★)

a How long does it take to roast a chicken that weighs 2 kg?

Give your answer in hours and minutes. (2 marks)

....

b What time should a 1.5 kg chicken be put in the oven for a Sunday lunch at 1.30 pm? (2 marks)

....

[Total: 4 marks]

(2) A locksmith charges £l to change the locks on a door, and then £k for each new key. (★★★)

a Write a formula for C, the cost of using the locksmith, ◄ if a customer orders n keys. (2 marks)

....

> Use different numbers for l, k and n to help you write a calculation. For example, a locksmith charges £50 to change the locks on a door, and then £5 for each new key. Then substitute the letters back in to write the formula.

b Use your formula to find C, in £, when $l = 90$, $k = 6.5$ and $n = 3$. (1 mark)

....

[Total: 3 marks]

(3) Rearrange each formula to make q the subject. (★★★★)

a $p = \dfrac{qs}{3}$ (2 marks)

....

> Decide what is happening to q, and then do the inverse to all terms on both sides of the equation.
> What is happening to q:
> $$\times s \div 3$$
> So do the inverse:
> $$\times 3 \div s$$

b $p = \dfrac{q}{r} + t$ (2 marks)

....

c $p = 3(q + r)$ (2 marks)

....

◄ First of all, multiply out the brackets.

d $p = \sqrt{2q}$ (2 marks)

....

◄ First square both sides.

[Total: 8 marks]

 SNAPIT! **Subject of a formula**

In a formula, the letter on its own on one side of the equals sign is called the subject.

Linear sequences

(1) The position-to-term rule for a sequence is multiply by 5 and add 1. (★★)

a Write the first four terms of the sequence. (2 marks)

> The term in position 1 is 1 × 5 + 1...
> The term in position 10 is 10 × 5 + 1

....

b What is the 50th term of this sequence? (1 mark)

....

[Total: 3 marks]

(2) Here are the first three patterns in a sequence.

The patterns are made from triangles and dots. (★★)

Pattern 1   Pattern 2   Pattern 3

a How many dots are there in Pattern 7? (2 marks)

....

b Rachel says, 'There are four triangles in Pattern 2.
  This means there will be eight triangles in Pattern 4.'

  Is Rachel correct? Give a reason for your answer. (2 marks)

> There are 2 marks for part b, so make 2 different statements.

....

....

[Total: 4 marks]

(3) Here are the first four terms of an arithmetic sequence. (★★★)

3   14   25   36

a Write down an expression for the *n*th term of the sequence. (2 marks)

....

SNAP IT! Finding the *n*th term

> To find the *n*th term, start by finding the common difference between the terms.
> The *n*th term will be (common difference)*n*, then + or − a whole number.

b Is 100 a term in this sequence? Show how you get your answer. (2 marks)

....

....

[Total: 4 marks]

Non-linear sequences

(1) Write the next two terms in each of these sequences. (★★)

a 1, 2, 4, 8 (1 mark)

....

b 1000, 100, 10, 1 (1 mark)

....

c 1.5, −3, 6, −12 (2 marks)

....

d What type of sequences are these? (2 marks)

....

[Total: 6 marks]

(2) This sequence follows the rule of a Fibonacci sequence. (★★★)

3, 3, 6, 9

a What is the next term? (1 mark)

....

b What is the position of the first term that is over 100? (2 marks)

....

[Total: 3 marks]

(3) Each day a gardener recorded the number of ladybirds on his herbs. (★★★)

Day	Mon	Tue	Wed	Thu	Fri
Number of ladybirds	2	8	32		

a He says, 'If the number of ladybirds continues to grow in the same way, by the end of Friday there will be more than 500 in my herb garden.'

> Complete the table to show the number of ladybirds on the next two days.

Is he correct? Show how you get your answer. (2 marks)

....

b If the sequence continues, on what day will there be more than 2000 ladybirds in the herb garden? (1 mark)

....

[Total: 3 marks]

(4) The nth term of a sequence is $\frac{1}{2}n^2$. (★★★★)

a Write down the first three terms of the sequence. (2 marks)

....

b Is 32 in the sequence? Show how you get your answer. (3 marks)

....

[Total: 5 marks]

Show that...

(1) Show that $2\left(x + \frac{1}{2}\right) \equiv x + x + 1$ (1 mark, ★★)

> Multiply the brackets and then simplify, to show the left-hand side and the right-hand side are the same.

(2) Show that $(x + 5)(x - 5) + 9 \equiv (x + 4)(x - 4)$ (2 marks, ★★★)

(3) Show that the sum of three consecutive numbers is a multiple of 3. (2 marks, ★★★★)

> Consecutive numbers follow each other in order. For example, 1, 2, 3 or 7, 8, 9... Notice that each consecutive number is one more than the previous number. This means you can write three consecutive numbers as $n, n + 1, n + 2$.

(4) Here is a path around a pond.

The path is made of concrete slabs. All the concrete slabs are the same size.

Each slab has a length of x cm and a width of y cm.

The perimeter of the pond is P cm. (★★★★★)

DOIT!

Write all the information you know on the diagram.

Write the distance between the corner and the end of a slab at the top and bottom of the pond, in terms of x and y:

a Show that $P = 16x - 4y$ (2 marks)

b Sanjit says that when x and y are whole numbers P is always a multiple of 4. Is Sanjit correct? Give a reason for your answer. (2 marks)

....

....

[Total: 4 marks]

Number machines

 SNAP IT! **Functions and rules**

A rule used in a number machine is called a function. The inverse function reverses the rule.

So, if a function is × 2 + 1; then the inverse function is − 1 ÷ 2.

(1) Here is a number machine. (★★)

x ⟶ ☐ × 4 ⟶ ☐ − 1 ⟶ y

> This number machine tells you that the rule is:
> take x, multiply by 4, subtract 1, and you get y.

a Work out y when x is 3. (1 mark)

....

b Work out x when y is 23. (2 marks) ◄— Use the inverse function to work out x.

....

c Write the equation in terms of x and y. (1 mark)

....

[Total: 4 marks]

(2) Complete the table for the number machine. (2 marks, ★★★)

x ⟶ ☐ × 2 ⟶ ☐ + 3 ⟶ y

x	y
−2	
0	
	9

(3) Complete the table for the function $y = \frac{x}{2} + 1$. (2 marks, ★★★★) ◄— First, draw a number machine.

x	y
−2	
1	
	5

Coordinates and midpoints

(1) Here is a graph showing the points *A* and *C*. (★★)

a Write down the coordinates of point *A*. (1 mark)

....

NAILIT!

1 Move your finger slowly along the horizontal (*x*) axis until the point is directly above or below your finger. This is the *x* coordinate.

2 Move your finger slowly up or down the vertical (*y*) axis until the point is directly left or right of your finger. This is the *y* coordinate.

SNAPIT! Coordinates

Coordinates are always written (*x*, *y*).

b On the graph, mark with a cross (×) the point (3, 1). Label it *B*. (1 mark)

c Write down the coordinates of point *C*. (1 mark)

....

[Total: 3 marks]

(2) Here is a graph showing the points *P*, *Q* and *R*. (★★★)

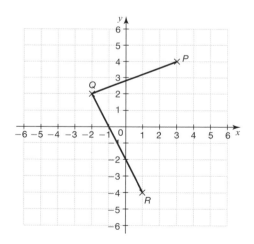

a Write down the coordinates of point *R*. (1 mark)

....

b On the graph, mark with a cross (×) a point *S*, so that the quadrilateral *PQRS* is a parallelogram. Join the points to show the parallelogram. (1 mark)

[Total: 2 marks]

(3) Here is a graph showing the triangle *X*, *Y* and *Z*. (★★★★)

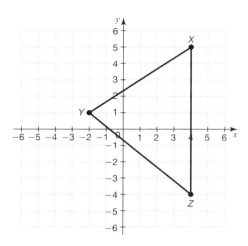

a Write down the midpoint of side *XY*. (2 marks)

> Add the *x* coordinates for point *X* and *Y*. Then halve it to find the midpoint of the *x* coordinate.
>
> Add the *y* coordinates for point *X* and *Y*. Then halve it to find the midpoint of the *y* coordinate.

....

b Write down the midpoint of side *XZ*. (2 marks)

....

c Write down the midpoint of side *YZ*. (2 marks)

....

[Total: 6 marks]

Straight-line graphs

(1) **a** Complete the table for $y = 2x + 3$ (2 marks, ★)

x	−1	0	1	2
y				

b Draw the graph of $y = 2x + 3$ (2 marks)

[Total: 4 marks]

(2) **a** Using the axes for graph C, draw the graph of $y = 3x$ for $x = -2$ to 2. Label it A. (2 marks, ★★)

> First of all, draw and complete a table with x from −2 to 2.

SNAP IT! Straight-line graph

The equation of a straight-line graph is $y = mx + c$, where m is the gradient (this is how far the graph rises or falls for every 1 unit across).

You can find the gradient by choosing any two points on a graph and working out: $\dfrac{\text{difference in } y \text{ coordinates}}{\text{difference in } x \text{ coordinates}}$.

c is the y-intercept (this is where the graph crosses the y-axis).

WORKIT!

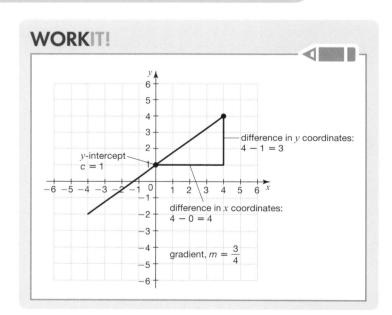

difference in y coordinates:
$4 - 1 = 3$

y-intercept $c = 1$

difference in x coordinates:
$4 - 0 = 4$

gradient, $m = \dfrac{3}{4}$

b Using the axes for graph C, draw the graph of $2x + y = -\dfrac{1}{2}$

Label it B. (2 marks, ★★★★)

> First, rearrange the equation to make y the subject.

c Work out the equation of the straight-line graph C. (2 marks, ★★★★)

....

[Total: 6 marks]

3 Here are the equations of three straight lines. (★★★★)

A $y = x + 1$ B $y = 2x + 1$ C $y = 2x + 2$

   a Which two lines are parallel? Give a reason for your answer. (2 marks)

   ....

   b Which two lines cross the y-axis at the same coordinates? Give a reason for your answer. (2 marks)

   ....

[Total: 4 marks]

4 A straight line passes through the points with coordinates (3, 2) and (−1, −6).

Work out the equation of the line. (3 marks, ★★★★★) ◄

> Work out the gradient.
> Substitute the gradient, along with $x = 3$ and $y = 2$, into the equation $y = mx + c$.
> Rearrange the equation to find c.
> Write the equation of the line.

....

SNAP IT!   More about straight-line graphs

Straight-line graphs
- are of the form $y = mx + c$
- are parallel if they have the same gradient, m
- cross the y-axis at the same coordinates if they have the same y-intercept, c.

5 Here is the graph of $y = 2x − 4$ (★)

   a Use the graph to find the value of y when x is 3. (1 mark)

   ....

> **1** Move your finger along the x-axis to 3.
> **2** Draw a straight line directly up from there to the graph.
> **3** Draw a straight line from this point on the graph directly left to the y-axis.
> **4** Read off the value of y.

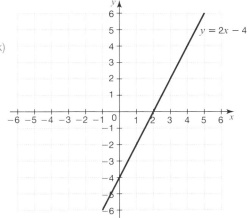

   b Use the graph to find the value of x when y is −2. (1 mark) ◄

   ....

[Total: 2 marks]

> **1** Move your finger along the y-axis to −2.
> **2** Draw a straight line directly right from this point to the graph.
> **3** Draw a straight line from this point on the graph directly up to the x-axis.
> **4** Read off the value of x.

6 Here is the graph of $2x + y = 3$ (★★★)

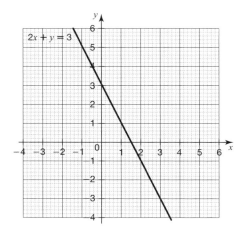

a Use the graph to find the value of x when y is -2. (1 mark)

....

b Use the graph to find the value of y when x is -0.5. (1 mark)

....

c Use the graph to find the value of y when x is 1.2. (1 mark)

....

[Total: 3 marks]

7 Here is the graph of $y = 3x - 1$ (★★★)

a On the same axis, draw the graph of $y = -3$ (1 mark)

On the line $y = -3$, every y coordinate has the value -3.

b Use the graph to estimate a solution to the equation $3x - 1 = -3$ (1 mark)

....

What is the value of x where the graphs intersect (cross)?

[Total: 2 marks]

8 Draw the graph of $y = 4x - 3$ for values of x from -5 to 5.

Use the graph to estimate a solution to the equation $4x - 3 = 2$ (3 marks, ★★★★★)

Solving simultaneous equations

① Solve algebraically these simultaneous equations. (★★★)

a $3x + y = 15$

$y = 2x$ (1 mark) ◄── Replace y with $2x$ in the first equation. That eliminates the y.

c $3x + y = 4$

$2x - y = 6$ (2 marks) ◄── Add the equations. That eliminates the y.

....

....

b $5x + y = 21$

$2x + y = 9$ (2 marks) ◄── Subtract the second equation from the first equation. That eliminates the y.

....

[Total: 5 marks]

WORKIT!

Solve the simultaneous equations

$2x + 3y = 5$

$3x + y = 11$

$2x + 3y = 5$ (1)

$3x + y = 11$ (2) ◄── Label the equations (1) and (2).

Equation (2) × 3 gives $9x + 3y = 33$ (3) ◄── You need to make the coefficients (the numbers before the letters) in x the same for both equations, or the coefficients in y the same for both equations.

Now we have:

$2x + 3y = 5$ (1)

$9x + 3y = 33$ (3) ◄── Now the coefficients in y are the same in equations (1) and (3).

Subtracting equation (1) from equation (3):

$7x = 28$

$x = 4$

Using equation (1), if $x = 4$, then: ◄── Substitute the value of x in any of the three equations to work out the value of y.

$8 + 3y = 5$

$3y = -3$

$y = -1$

The solution is $x = 4$, $y = -1$

(2) Solve algebraically these simultaneous equations. (★★★★)

a $2x + 2y = 14$

$3x + y = 11$ (3 marks)

b $4x - 2y = 2$

$2x - 3y = 7$ (3 marks)

c $2x + 3y = 20$

$3x + 2y = 15$ (4 marks)

....

[Total: 10 marks]

NAILIT!

Sometimes it is necessary to multiply both the first and the second equation by a number so that the coefficients in x or the coefficients in y in both equations are the same.

(3) **a** Complete this table for $x + y = 2$ Complete this table for $2x - y = 1$ (2 marks, ★★★★)

x	0	
y		0

x	0	
y		0

b Solve graphically these simultaneous equations. (1 mark, ★★★★★)

$x + y = 2$

$2x - y = 1$

NAILIT!

To solve simultaneous equations graphically, draw the graph of each equation and find the values of x and y where they intersect (cross).

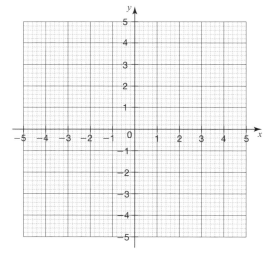

c Use the same axes to solve graphically these simultaneous equations. (1 mark, ★★★★★)

$2x - y = 1$

$-x + 2y = 4$

[Total: 4 marks]

Quadratic graphs

① Here are five graphs. (★★)

a Which of these graphs are **not** quadratic graphs? (1 mark)

....

b Which graph shows the equation $y = x^2$? (1 mark)

....

c Which graph shows the equation $y = -x^2$? (1 mark)

....

d Which graph shows the equation $y = 1$? (1 mark)

....

e Which graph shows the equation $y = x^2 + 1$? (1 mark)

....

[Total: 5 marks]

② **a** Complete the table of values for $y = x^2 - 3$. (2 marks, ★★★)

x	−3	−2	−1	0	1	2	3
y							

b Draw the graph of $y = x^2 - 3$ for values of x from −3 to 3 (3 marks, ★★★)

c Write down the equation of the line of symmetry. (1 mark, ★★★)

.... ◄── The line of symmetry is at $x =$ ☐

d Write down the coordinates of the turning point. (1 mark, ★★★)

....

[Total: 7 marks]

NAILIT!

The turning point is where the graph turns.

NAILIT!

When drawing quadratic graphs
- use a sharp pencil
- plot each point with a cross (×)
- draw a smooth curve through **all** the points.

3 **a** Draw the graph of $y = x^2 + 2x - 1$ for values of x from -3 to 1. (3 marks, ★★★★)

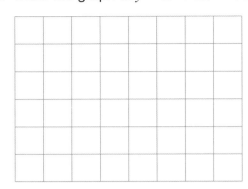

b Write down the equation of the line of symmetry. (1 mark, ★★★★)

....

c Write down the coordinates of the turning point. (1 mark, ★★★)

....

[Total: 5 marks]

4 Here is the graph of $y = x^2 + x + 3$. (★★★)

a On the same axis draw the graph of $y = 5$. (1 mark)

b Use the graphs to find solutions to the equation
$x^2 + x + 3 = 5$. (2 marks)

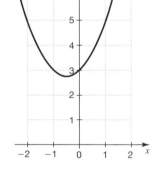

....

Work out the values of x where the graphs intersect (cross). There are two places were the graphs intersect, so there are two solutions to this equation.

[Total: 3 marks]

5 Here is the graph of $y = x^2 + 2x - 3$.

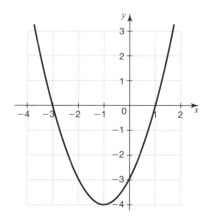

Use the graph to find solutions to the equation $x^2 + 2x - 3 = 0$.

(2 marks, ★★★)

Look at where the graph intersects with $y = 0$.

....

6 **a** Draw the graph of $y = x^2 + 3x + 1$ for values of x from -3 to 0. (3 marks, ★★★★★)

Draw and complete a table of values of $y = x^2 + 3x + 1$ for values of x from -3 to 0.

x	-3	-2	-1	0
y				

b Use the graph to find solutions to the equation $x^2 + 3x + 1 = -1$. (2 marks, ★★★★★)

....

c Use the graph to estimate solutions to the equation $x^2 + 3x + 1 = 0$. (2 marks, ★★★★★)

....

[Total: 7 marks]

Solving quadratic equations

(1) Solve (★★)

a $x^2 + 6x = 0$ (2 marks) ◄————

> Factorise using brackets.
> Whatever is outside the bracket multiplied by whatever is inside the bracket = 0.
> This means, either whatever is outside the bracket = 0 or whatever is inside the bracket = 0
> Now you can give two possible solutions for x.

....

b $y^2 - 11y = 0$ (2 marks)

c $3d^2 - 9d = 0$ (2 marks)

....

[Total: 6 marks]

(2) Solve (★★★★)

a $x^2 - 16 = 0$ (1 mark)

> **NAILIT!**
>
> Whenever a quadratic is $x^2 -$ a square number, then it can be factorised as $(x + \quad)(x - \quad)$.
>
> square root of square number

....

b $a^2 - 81 = 0$ (1 mark)

c $z^2 - 100 = 0$ (2 marks)

....

[Total: 4 marks]

(3) Solve (★★★★)

a $x^2 + 5x + 6 = 0$ (2 marks) ◄————

> **1** Begin to factorise using brackets, by writing $(x \quad)(x \quad) = 0$
> **2** Write all the factor pairs that make 6, including any negative ones. For example: 1, 6; −1, −6
> **3** Which factor pair adds to the number in the x term, 5?

....

b $x^2 + 3x - 10 = 0$ (3 marks)

c $x^2 - 9x + 14 = 0$ (3 marks)

....

[Total: 8 marks]

(4) Work out the roots of these functions. (★★★★★)

a $y = x^2 - 3x$ (2 marks) ◄——

> Write y as 0. Now solve as a quadratic equation.

> **NAILIT!**
>
> The roots of a quadratic function are the solutions when $y = 0$.

....

b $y = x^2 - 25$ (2 marks)

c $y = x^2 + 3x - 18$ (2 marks)

....

[Total: 6 marks]

Cubic and reciprocal graphs

(1) Here are five graphs. (★)

a Which of these graphs are **not** cubic graphs? (1 mark)

....

b Which graph shows the equation $y = x^3$? (1 mark)

....

c Which graph shows the equation $y = -x^3 + 1$? (1 mark)

....

d Which graph shows the equation $y = \frac{1}{x}$? (1 mark)

....

[Total: 4 marks]

(2) a Complete the table of values for $y = x^3 + 4$. (2 marks, ★★★)

x	−2	−1	0	1	2
y					

b Draw the graph of $y = x^3 + 4$ for values of x from −2 to 2. (3 marks, ★★★★)

 NAIL**IT!**

When drawing cubic graphs

- use a sharp pencil
- plot each point with a cross (×)
- draw a smooth curve through **all** the points.

[Total: 5 marks]

③ Here is the graph of $y = x^3 - 8$ (★★★★)

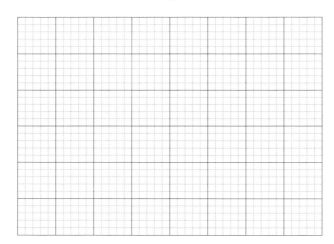

a What type of graph is it? (1 mark)

....

b Write the coordinates where the curve crosses the y-axis. (2 marks)

....

c Write the coordinates where the curve crosses the x-axis. (2 marks)

....

[Total: 5 marks]

④ a Complete the table of values for $y = \frac{1}{x} + 1$. (3 marks, ★★★★★)

x	-3	-2	-1	$-\frac{1}{2}$	$\frac{1}{2}$	1	2	3
y	$\frac{2}{3}$			-1				

b Draw the graph of $y = \frac{1}{x} + 1$ for values of x from -3 to 3. (3 marks, ★★★★★)

[Total: 6 marks]

Drawing and interpreting real-life graphs

(1) The graph shows the cost of a taxi in New York, in US dollars. (★★★★)

a How much is the initial charge? (1 mark)

....

b How much is the charge per mile? (2 marks)

> The gradient of the graph,
> $\dfrac{\text{difference in } y \text{ coordinates}}{\text{difference in } x \text{ coordinates}}$
> gives the charge per mile.

....

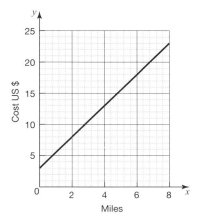

c You can use the information in the table to convert between UK pounds and US dollars.

UK £	0	5	20	50
US $	0	6	24	60

Use this information to draw a conversion graph. (2 marks)

d How far can you travel in a New York taxi for £15? (2 marks)

> You will need to use both graphs.

....

[Total: 7 marks]

(2) The graph shows the height of a ball, in metres, against time, in seconds. (★★★)

a What is the maximum height the ball reaches? (1 mark)

....

b At how many seconds does the ball reach its maximum height? (1 mark)

....

c After how many seconds does the ball hit the ground? (1 mark)

....

d There were two times when the ball was at a height of 12 m. Write down these times. (2 marks)

....

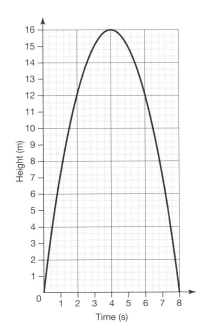

[Total: 5 marks]

③ The graph shows a short part of a cyclist's cycle ride. (★★★★)

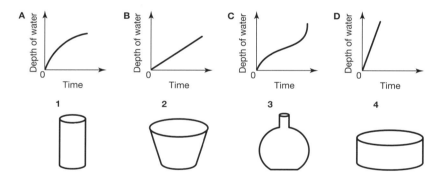

a What is the cyclist's maximum speed? (1 mark)

....

b Describe the cyclist's ride between 5 and 20 seconds. (2 marks)

....

c What is the cyclist's acceleration for the final 20 seconds of the ride? (2 marks) ◄—

> The gradient of the graph
>
> $$\frac{\text{difference in } y \text{ coordinates}}{\text{difference in } x \text{ coordinates}}$$
>
> gives the cyclist's acceleration in m/s².

....

d What happens after 40 seconds? (1 mark)

....

[Total: 6 marks]

④ a Match each graph to a container (3 marks, ★★★★★)

A Depth of water / Time
B Depth of water / Time
C Depth of water / Time
D Depth of water / Time

1   2   3   4

....
....
....
....

b Sketch a depth–time graph for this container. (3 marks, ★★★★★)

[Total: 6 marks]

Ratio, proportion and rates of change
Units of measure

1 Convert these measurements. (★)

a 4 m to cm (1 mark)

> **Ask yourself:**
>
> **1** What is the conversion factor? (How many cm in 1 m?)
>
> **2** Do you expect the number in your answer (in cm) to be more or less than the original number (4)?
>
> **3** If more, then multiply by the conversion factor; if less, then divide by the conversion factor.

....

b 5000 g to kg (1 mark)

....

c 1.5 litres to ml (1 mark)

d 8250 m to km (1 mark)

....

[Total: 4 marks]

2 Sally made 6 litres of lemonade. Her friends drank 3500 ml. How many ml did Sally have left? (2 marks, ★★★)

....

> **NAILIT!**
>
> Watch out for questions that have two different units of measure (for example litres and millilitres).
>
> Convert one of the measures so you are working in only one type of unit.

3 Luke and his brother Adam left home at the same time.

Luke ran to school in 240 seconds.

Adam ran to school in 3 minutes 47 seconds. (★★★)

a Which brother arrived at school first?

You must show your working. (2 marks)

....

b How long did the first boy at school wait for his brother to arrive? (1 mark)

....

[Total: 3 marks]

4 1 m ≈ 3.2 feet

Ben is 1.25 m; Tom is 4.8 feet. Who is taller?

You must show working to justify your answer. (2 marks, ★★★★)

....

Ratio

1. A child owns 20 books. 12 contain stories and the rest are for colouring. Write the ratio of story to colouring books in its simplest form. (2 marks, ★)

> Write the ratio ▇ : ▇
>
> Simplify by dividing both numbers by their highest common factor.

....

2. Holly has £15. She wants to save and spend it in the ratio 1 : 2.

   How much will she spend? (2 marks, ★★)

....

3. Chen has £60.

   He wants to save, spend and give some to charity in the ratio 1 : 2 : 3.

   How much will he give to charity? (2 marks, ★★★)

....

4. Phil is making green paint.

   He plans to mix blue and yellow paint in the ratio 3 : 7 to make 5 litres of green paint.

   He has 2 litres of blue paint and 3 litres of yellow paint.

   Does he have enough blue and yellow paint to make the green paint?

   Show working to justify your answer. (4 marks, ★★★★)

DO IT!

Draw a diagram of the ratio.
All boxes = 5 litres = 5000 ml.
1 box = ▢ ml
3 boxes represent blue paint. How much blue paint does Phil need? Does he have enough blue paint?
7 boxes represent yellow paint. How much yellow paint does Phil need? Does he have enough yellow paint?

....

Scale diagrams and maps

(1) Complete these sentences. (1 mark, ★)

A scale of 1 : 10 000 on a map means

1 cm on the map is _____ cm in real life.

This means 1 cm on the map is _____ m in real life.

(2) On a map of a village 1 cm represents 50 m.

In real life, the village shop is 150 m from the bus stop.

How far is this, in centimetres, on the map? (1 mark, ★★)

....

(3) The diagram below represents two trees in a field.

Tree A   Tree B
  ×   ×

Scale: 1 cm represents 4 m.

Work out the distance, in metres, between tree A and tree B. (2 marks, ★★★)

....

(4) A scale model of Big Ben in London is 1 : 400

Big Ben is 96 m high.

How tall is the model? (2 marks, ★★★★)

....

Write

1 cm on the model is ☐ cm in real life.

1 cm on the model is ☐ m in real life.

☐ cm on the model is 96 m in real life.

Fractions, percentages and proportion

(1) Ali and Bess share a pie in the ratio 1 : 3.

What fraction does Bess receive? (2 marks, ★★)

....

SNAPIT! **Diagram for ratios**

Draw a diagram for ratios. It will help you convert them to fractions.

For example: A : B = 1 : 3 | A | B | B | B |

(2) In a shopping basket, 1 item is a fish, 3 items are fruit, 6 items are tins. (★★)

   a Write the items in the basket as a ratio of fish : fruit : tins. (1 mark)

     ....

NAILIT!

To write one measure as a fraction of another, both measures must be in the same units.

   b What fraction of items in the basket is fruit? (1 mark)

     ....

   c What percentage of items in the basket are tins? (1 mark)

SNAPIT! **Percentage**

To write one quantity as a percentage of another:
1 Write as a fraction.
2 Multiply by 100 to convert to a percentage.

     ....

[Total: 3 marks]

(3) What is 50 g as a fraction of 4 kg? (1 mark, ★★)

....

(4) A dance school has street dancers, ballet dancers and tap dancers in the ratio 3 : 8 : 14

What percentage of the dance school do ballet dance? (2 marks, ★★★★)

....

Direct proportion

1 Five concert tickets cost £80. (★)

a How much will one concert ticket cost? (1 mark)

....

b How much will nine concert tickets cost? (1 mark)

....

[Total: 2 marks]

2 This graph shows the cost of buying packs of 10 colouring pencils. (★)

a What is the cost of 6 packs of 10 pencils?

Give your answer in £. (1 mark)

....

b What is the cost of 1 pencil? (1 mark)

....

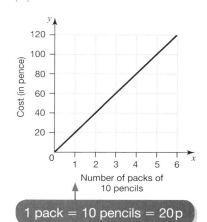

1 pack = 10 pencils = 20p

c State **two** ways this graph shows that packs of pencils are in direct proportion to the cost. (2 marks)

....

....

[Total: 4 marks]

3 A curry recipe is for 4 people. Here are the spices it asks for. (★★★)

1 teaspoon of turmeric
2 teaspoons of chilli powder
$2\frac{1}{2}$ teaspoons of cumin

a Sally is using the recipe to make curry for her class. There are 28 people in Sally's class. How many teaspoons of each spice will she need? (3 marks)

....

....

....

b 1 teaspoon of chilli powder ≈ 3g
Sally has 75g of chilli powder. Does she have enough to make the curry for her class?
Show working to justify your answer. (3 marks)

....

[Total: 6 marks]

Inverse proportion

(1) This graph shows competition prize money divided by different numbers of winners. (★)

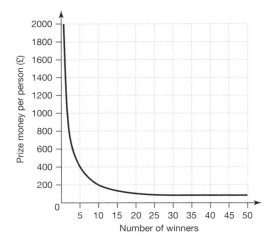

a   If there are 5 winners, how much money will they each get? (1 mark)

....

b   A winner receives £200. How many other winners are there? (1 mark)

....

c   How much is the prize money in total? (1 mark)

....

d   Describe the relationship between prize money and number of winners. (1 mark)

....

[Total: 4 marks]

SNAPIT!   Direct proportion

If two values are in direct proportion, as one increases, the other increases at the same rate.

For example, if one value increases by doubling (× 2), then the other value increases by doubling (× 2).

If two values are in inverse proportion, then as one increases, the other decreases at the same rate.

For example, if one value increases by doubling (× 2), then the other value decreases by halving (÷ 2).

(2) It takes 3 people 2 hours to decorate a room for a party. (★★)

   a How long would it take 1 person? (1 mark) ◄———

> Would it take 1 person a longer or shorter amount of time?

....

   b How long would it take 12 people? (2 marks) ◄———

> It is difficult to work with 3 and 2, as they don't divide nicely into each other. Try changing 2 hours into 120 minutes.

[Total: 3 marks]

....

(3) $4y = \frac{4}{x}$ (★★)

   a Describe what happens to the value of y as the value of x increases from 1 to 8. (1 mark)

....

   b Is x directly proportional or inversely proportional to y? Explain why. (1 mark)

....

[Total: 2 marks]

(4) It takes a printer 4 minutes to print a 240-page document.

How long will it take the printer to print a 600-page document? (2 marks, ★★★)

....

Working with percentages

(1) In a sale, the price of a coat was reduced from £125 to £75.

Work out the percentage decrease. (1 mark, ★★★)

SNAPIT! **Percentage increase or decrease**

To work out percentage increase or decrease:
1 Work out the amount of increase or decrease.
2 Divide by the original value.
3 Multiply by 100.

....

(2) The table shows sales for two different companies in 2006 and 2016. (★★★)

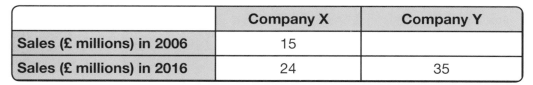

	Company X	Company Y
Sales (£ millions) in 2006	15	
Sales (£ millions) in 2016	24	35

a Work out the percentage increase in sales for Company X between 2006 and 2016. (1 mark)

Decide which numbers in the table you need to find your answer.

....

b Company Y achieved a percentage increase in their sales of 25% between 2006 and 2016.

Work out their sales (£ millions) in 2006. (1 mark)

You know that 125% of the original value in 2006 is £35 million. Use this to work out 1%, then 100%.

....

[Total: 2 marks]

(3) Sarah invests £265 in a savings account. It pays 2% interest. (★★★)

a How much will Sarah have in her account after 3 years if the account pays simple interest. (2 marks)

....

b How much will Sarah have in her account after 3 years if the account pays compound interest? (3 marks)

For simple interest: work out how much interest is received after 1 year. Multiply this by 3 and add to £265. For compound interest: work out how much Sarah has after 1 year. Use this amount to find out how much she has after 2 years, and so on.

....

[Total: 5 marks]

Compound units

(1) Convert 3 km/minute to metres per second. (1 mark, ★★)

> Work with each unit separately.
> Convert 3 km to m: 3 km/minute = ▢ metres/minute
> If that's the speed in 1 minute. How far is that in 1 second?

....

(2) A tank holds 20 litres of water. Water is poured into the tank at a rate of 5 litres per minute. How long does it take to fill the tank? (1 mark, ★★★)

 STRETCHIT!

> Work out the same calculation in gallons, given that
> 1 litre = 4.546 gallons.

....

(3) Pressure = $\frac{\text{force}}{\text{area}}$

Work out the pressure exerted by a force of 300 Newtons on an area of 0.05 m².

Write your answer using the unit Newtons/m². (2 marks, ★★★)

....

(4) On Saturday, Sami drives from London to Leeds. It takes her 4 hours.

Her average speed is 50 miles per hour.

On Sunday, Sami drives from Leeds to Glasgow. It is 356 km.

5 miles ≈ 8 kilometres

On which day does Sami drive furthest?
Show working to justify your answer. (4 marks, ★★★★★)

> Sometimes questions don't tell you exactly what to do. Read carefully what the question is asking – this one asks on which day Sami drives furthest. Then ask yourself, what you need to find out to answer it. For this question, you need to find the distance Sami drove on Saturday, and the distance she drove on Sunday.

....

Geometry and measures
Measuring and drawing angles

(1) Measure these angles. (★)

a

.... (1 mark)

STRETCHIT!

The diagram in 1a shows an obtuse angle. If you marked the other angle around the same point (360° − *a*) it would be a reflex angle. Can you draw a quadrilateral, with one line of symmetry, containing a reflex angle?

b

Estimate the size of the angles first. That will give you confidence that you have measured them correctly.

.... (1 mark)

c

.... (1 mark)

[Total: 3 marks]

(2) Draw and label angle $x = 193°$ (2 marks, ★★) ← How much bigger than 180° (a straight line) will your angle be?

(3) Alfie is thinking of an angle. (★★★)

a He says it is obtuse, and its size is a multiple of 20.

Write down all possible sizes for Alfie's angle. (2 marks)

....

b Clare is thinking of two angles. She says her first angle plus her second angle gives an answer of 87°.

What type of angles is she thinking of? Give a reason for your answer. (2 marks)

....

....

[Total: 4 marks]

Using the properties of angles

SNAP IT! Properties of angles

Angles on a straight line always add up to 180°.

Angles in a triangle always add up to 180°.

Quadrilaterals can always be divided into two triangles. Therefore, angles in a quadrilateral always add up to $2 \times 180° = 360°$

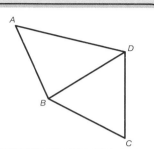

(1) Work out the size of the angles marked x in the diagrams.

You must show your working. (★)

a A (2 marks)

b A   B (2 marks)

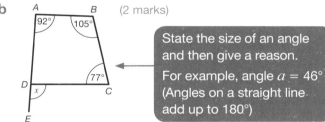

State the size of an angle and then give a reason.

For example, angle $a = 46°$ (Angles on a straight line add up to 180°)

....

[Total: 4 marks]

(2) Work out the size of the angles marked x in the diagrams.

You must show your working. (★★★)

a (2 marks)

b (2 marks)

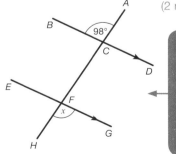

Look for equal angles. For example, base angles of an isosceles triangle are equal; corresponding angles are equal; vertically opposite angles are equal; alternate angles are equal.

....

[Total: 4 marks]

(3) Work out the size of the angle marked x in the diagram.

You must show your working. (2 marks, ★★★★)

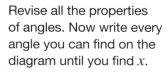

....

(4) Show that this is an equilateral triangle. (3 marks, ★★★★★)

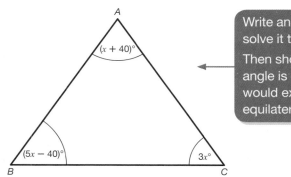

Write an equation and solve it to find x.

Then show that each angle is exactly as you would expect in an equilateral triangle.

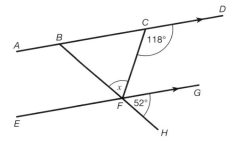

NAILIT!

Revise all the properties of angles. Now write every angle you can find on the diagram until you find x.

STRETCHIT!

A triangle has sides of 10, x and x cm. It has angles of 60°, $\theta°$ and $\theta°$.

Can you say what kind of triangle it is?

Using the properties of polygons

1 **a** What do the interior angles of a hexagon add up to? (2 marks, ★★)

....

b Work out the size of an interior angle of a regular hexagon. (1 mark, ★★)

....

c Work out the size of an exterior angle of a regular hexagon. (1 mark, ★★)

....

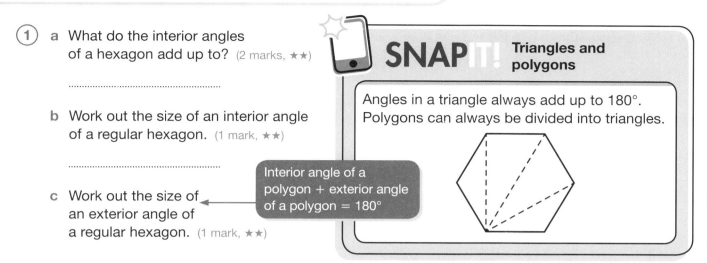

SNAP IT! **Triangles and polygons**

Angles in a triangle always add up to 180°.
Polygons can always be divided into triangles.

Interior angle of a polygon + exterior angle of a polygon = 180°

[Total: 4 marks]

SNAP IT! **Angles in a polygon**

There are always two fewer triangles than the number of sides.
Therefore, angles in a polygon always add up to
(number of sides − 2) × 180°.

If you bring together all the exterior angles, they make the angles around a point.
Therefore, the sum of the exterior angles = 360°.

Interior angle

Exterior angle

2 Here is a polygon. (★★★)

a Name the polygon. (1 mark)

....

b It is a regular polygon. Write **two** things you know about it. (2 marks)

....

c Work out the size of an interior angle, *a*. (1 mark)

[Total: 4 marks]

....

3 The interior angle of a regular polygon is 144°. Name the polygon. Show working to justify your answer. (3 marks, ★★★★)

First of all find the size of an exterior angle.
What do you know about the sum of all the exterior angles?

Using bearings

① A ship sails from point *A* to point *B* to point *C*. (★★★)

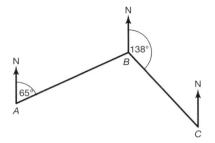

a Work out the bearing of *B* from *A*. (1 mark)

....

The North arrows are parallel.

Therefore, use co-interior angles to find the angle between the line *BC* and the North arrow at *C*.

Then use angles at a point to find the bearing.

b Work out the bearing of *B* from *C*. (1 mark)

....

c Work out the bearing of *A* from *B*. (2 marks)

....

[Total: 4 marks]

② The bearing of *X* from *O* is 276°. Work out the bearing of *O* from *X*. (2 marks, ★★★★)

Draw a diagram.

....

③ Here are points *P* and *Q*. (★★★★)

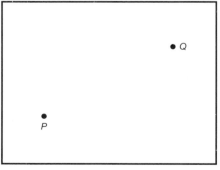

Draw a North line at *P*. Join the points *P* and *Q*. Put the base line of your protractor on North. Measure from 0° to *Q*.

a Measure the bearing of *Q* from *P*. (1 mark)

....

b Draw an accurate bearing of 093° from *P*. Draw an accurate bearing of 225° from *Q*. Mark the point where they intersect *X*. (2 marks)

[Total: 3 marks]

Properties of 2D shapes

① Here is part of a shape with one line of symmetry. (★★)

   a Complete the shape. (1 mark)

   b Name the shape. (1 mark)

     ....

   c Write one property of the shape that does not involve symmetry. (1 mark)

     ....

                              [Total: 3 marks]

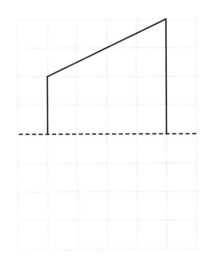

② Here is a quadrilateral. (★★)

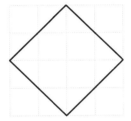

   a Draw any lines of symmetry. (2 marks)

   b What is its order of rotational symmetry? (1 mark)

     ....

   c Name the shape. (1 mark)

     ....

   d Write two properties of the shape that do not involve symmetry. (2 marks)

     ....

                              [Total: 6 marks]

NAILIT!

Properties of a shape are facts about its angles, sides, diagonals and symmetry.

NAILIT!

A quadrilateral has four straight sides.

③ a Tick the quadrilaterals that have two lines of symmetry. (2 marks, ★★★★★)

     ☐ rectangle   ☐ parallelogram   ☐ square   ☐ rhombus   ☐ kite

   b Sketch the quadrilaterals and mark the lines of symmetry on them. (6 marks, ★★★) ◄——

On your sketch, make sure you mark
- parallel sides
- equal sides
- right angles
- equal angles.

                              [Total: 8 marks]

Congruent shapes

① **Which shapes are congruent to shape A?**

Circle your answers. (1 mark, ★)

A   B   C   D   E   F

② These two triangles are congruent.

What size is angle x? (1 mark, ★★★) ◀

Sketch the second triangle, rotated, so that it is the same way round as the first triangle. Write any information from the first triangle on the second triangle.

....

③ Each pair of triangles is congruent. Explain why. (★★★★)

a   (1 mark)   **b**   4 cm   (1 mark)

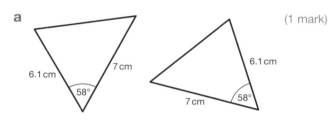

....

....

....

....

[Total: 2 marks]

④ **Are these two triangles congruent? Give a reason for your answer.** (2 marks, ★★★★★)

....

....

Constructions

NAILIT!

When drawing constructions in maths, it is important to

- use a sharp pencil
- label any points, lengths or angles you are given
- *never* rub out your construction arcs and lines.

(1) **a** Use a ruler and compasses to construct an accurate drawing of triangle *ABC*. (3 marks, ★★★)

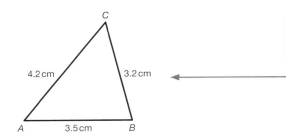

Use a ruler to draw a straight line *AB*, of 3.5 cm.
Open your compasses to 4.2 cm.
Put the point on *A*, and draw an arc above your line.
Open your compasses to 3.2 cm.
Put the point on *B*, and draw an arc above your line (which intersects your other arc).
Label the point where the arcs intersect, *C*.
Join *AB* and *BC* to finish constructing the triangle.

The bisector cuts the angle exactly in half.
Open your compasses so that they are a little less than the lengths of the arms (*AC* or *AB*) of the angle *BAC*.
Keep them exactly like this throughout your construction.
Put the point of your compasses on *A*.
Draw two arcs – one on *AB* and one on *AC*.
Put the point of your compasses on the arc on *AB*.
Draw an arc in between the arms of the angle *BAC*.
Put the point of your compasses on the arc on *AC*.

b Construct a bisector of angle *BAC*.
(3 marks, ★★★)

Draw an arc in between the arms of the angle *BAC* (which intersects your other arc).
Join the point where the arcs intersect to *A*.

[Total: 6 marks]

(2) **a** Use a ruler and compasses to construct an accurate drawing of triangle *XYZ*, where *XY* = 2.8 cm, *XZ* = 3.6 cm, *YZ* = 3 cm. (3 marks, ★★★★)

Sketch and label the triangle first. Then you can see the triangle you are aiming to construct.

b Construct the bisector of angle *XYZ*. (3 marks, ★★★★)

[Total: 6 marks]

3 **a** A person is standing at point *O*, in a field. He can see a straight road, *PQ*, ahead of him. (3 marks, ★★★★)

Use a ruler and compasses to construct the shortest distance from point *O* to the straight road.

The shortest distance is the perpendicular bisector.
Put the point of your compasses on *O*.
Draw two arcs on *PQ*.
Keep the compasses open, exactly as they are.
Put the point of your compasses on one arc on *PQ*.
Draw an arc below the line *PQ*.
Put the point of your compasses on the other arc on *PQ*.
Draw an arc below the line *PQ* (which intersects your other arc).
Join the point where the arcs intersect to *O*.

b The scale of this diagram is 1:100. Measure and write down the shortest distance. Give your answer in metres. (2 marks, ★★★★)

....

[Total: 5 marks]

STRETCHIT!

The Romans are famous for constructing straight roads across Britain. How do you think they made sure the roads were straight?

Sometimes Roman roads do change direction (they turn at an angle). Think about why that might be. (You could look up Fosse Way or Watling Street on the internet.)

Drawing circles and parts of circles

(1) **a** Draw a circle with diameter 7 cm. (1 mark, ★★) ◄─── You need to work out the radius to draw the circle.

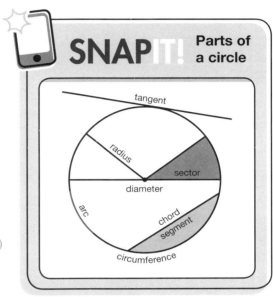

SNAPIT! **Parts of a circle**

tangent
radius
sector
diameter
arc
chord
segment
circumference

b Draw and label a tangent to the circle of length 4 cm. (1 mark, ★★)

c Shade and label a segment of the circle. (1 mark, ★★)

d Label a chord in the circle. (1 mark, ★★)

[Total: 4 marks]

(2) **a** Draw a sector of a circle with radius 3 cm and angle 42°. (2 marks, ★★★)

b Label the arc. (1 mark, ★★★)

[Total: 3 marks]

(3) Donald says, 'A segment of a circle and a sector of a circle are the same'.

Is Donald correct? You must give a reason for your answer. (3 marks, ★★★★)

....

....

....

Loci

(1) **a** Shade the locus of points less than 2.3 cm from point *A*. (2 marks, ★★★)

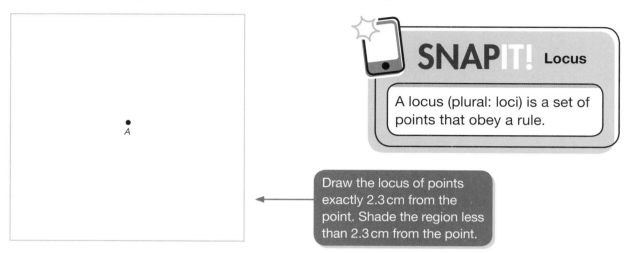

SNAPIT! Locus

A locus (plural: loci) is a set of points that obey a rule.

Draw the locus of points exactly 2.3 cm from the point. Shade the region less than 2.3 cm from the point.

b Draw the locus of points exactly 2.3 cm from line *BC*. (1 mark, ★★★)

B   C

c Draw the locus of points equidistant from *DE* and *EF*. (2 marks, ★★★)

Equidistant means the same distance.

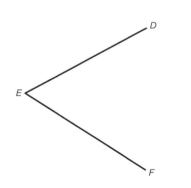

SNAPIT! Bisector

The locus of points equidistant from two lines that meet to make an angle is the bisector of the angle.

Don't forget to leave in any construction arcs and lines you use when drawing loci.

[Total: 5 marks]

(2) A path is laid so that it is exactly the same distance between a tree and a barn.

Draw the path. (2 marks, ★★★)

•
Barn

•
Tree

Imagine two points, A and B, joined by a line. The locus of points exactly the same distance from A and B is the perpendicular bisector of AB.

1:200 means 1 cm on the scale diagram represents 200 cm = 2 m in real life

(3) This square field *ABCD* is drawn to scale: 1:200

A dog and a goat are tethered in the field. The dog is tethered at *A* with a 4 m length of rope. The goat is tethered to a piece of wood nailed between two poles, at *X* and *Y*, with a 3 m length of rope. The rope can slide along the wood, between the poles.

Complete the scale drawing and shade the area that neither the dog, nor the goat, can reach. (3 marks, ★★★★★)

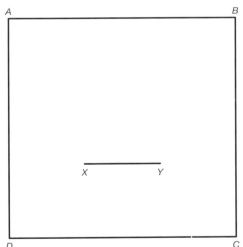

(4) Two cows graze in a field of length 20 m and width 12 m. They are tethered to diagonally opposite corners by ropes of length 13 m. On a scale drawing, show the area grazed by **both** cows. (3 marks, ★★★★)

Perimeter

(1) Work out the perimeter of a regular hexagon with sides 9 cm. (1 mark, ★)

....

(2) Work out the perimeter of this F shape. Give your answer in cm. (2 marks, ★★★)

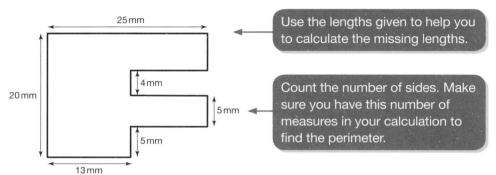

> Use the lengths given to help you to calculate the missing lengths.

> Count the number of sides. Make sure you have this number of measures in your calculation to find the perimeter.

....

(3) Greta has 1.5 m of lace. She wants to put it all the way around the outside of this seat cushion.

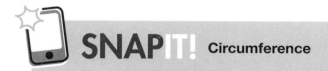

Does Greta have enough lace? You must show your working. (3 marks, ★★★★)

....

SNAPIT! **Circumference**

The perimeter of a circle is called the circumference.

The circumference of a circle = $2\pi r = \pi d$ r = radius d = diameter

Area

(1) What are the areas of these shapes? (★★)

a (1 mark)   **b** (1 mark)   **c** (1 mark)

Split into two shapes.

....

[Total: 3 marks]

SNAP IT!   Properties of shapes

The dashes on the sides of the shapes show these sides are parallel.

The angles marked with a ☐ show right angles (90°).

Area of a parallelogram = base length × perpendicular height (this is the height at right angles to the base).

Area of a trapezium = $\frac{1}{2}$ (parallel sides added together) × perpendicular height (this is the height at right angles to the parallel sides).

(2) One side of a rectangle is 6 cm. Its perimeter is 16 cm. Work out its area. (2 marks, ★★)

Often it can help to draw a diagram. Draw a diagram of the rectangle and put all the information you know on it. Can you use this information to work out other information about the rectangle? Does this help you to work out the area?

....

(3) What is the area of the shaded part of this shape? Give your answer to 2 d.p.

 You must show your working. (3 marks, ★★★★★)

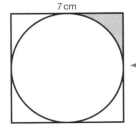

Begin with finding the area of the biggest shape (the square). Then subtract the area of the circle. What are you left with?

Sectors

(1) Here is a circle. (★★)

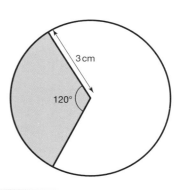

  a What fraction of the circle is grey? (1 mark)

....

  b What is the area of the grey sector?
    Give your answer in terms of π. (1 mark)

....

> Leaving π in your answer means that it is exact. Work out the fraction × r^2 and then write π.

SNAPIT! **Area of a sector**

Area of a sector = fraction of circle × πr^2

[Total: 2 marks]

(2) Here is a quarter of a circle. (★★★)

  a Work out its area. Give your answer to 1 d.p. (1 mark)

2.8 cm

....

NAILIT!

The perimeter is all the way around the shape. It includes the arc and two sides.

  b Work out its perimeter. Give your answer to 1 d.p. (2 marks)

....

[Total: 3 marks]

(3) The diagram shows a sector of a circle of radius 5 cm. (★★★★)

  a Work out the area of the sector. Give your answer to 2 d.p. (2 marks)

....

  b Work out the length of arc *AB*.
    Give your answer to 2 d.p. (2 marks)

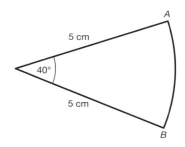

....

[Total: 4 marks]

3D shapes

1 **a** Sketch the net of this 3D shape. (1 mark, ★★)

b Name the shape. (1 mark, ★★)

....

c Complete this table for the shape. (3 marks, ★★)

SNAP IT! **Vertices**

A vertex is a corner. Vertices is the plural of vertex.

Number of faces	Number of edges	Number of vertices

[Total: 5 marks]

2 The diagram shows a 3D shape made of seven identical cubes. (★★)

On the grid, draw the

a plan view (1 mark)

b front view (1 mark)

c left side view. (1 mark)

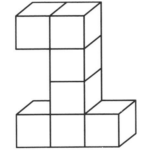

The plan view is the view from above.
The front view shows the shape from the front.
The side view shows the shape from the side.

[Total: 3 marks]

3 Here are the plan, front and side elevations of a 3D shape. (★★★★)

a Sketch the 3D shape. (3 marks)

Draw the front view, angled to the left (each square will become a rhombus).

Draw the side view, angled to the right.

Use the plan view, and what you have already drawn, to work out how to draw the top of the shape.

Plan

Front

Side

b How many identical cubes make the shape? (1 mark)

....

[Total: 4 marks]

Volume

(1) This prism is made from centimetre cubes.
Work out its volume. (1 mark, ★)

....

SNAP IT! Cross-section of prism

A prism has a cross-section shape that is the same all along its length.

This prism has this cross-section.

SNAP IT! Volume of prism

Volume of a prism = area of cross-section × length

NAIL IT!

There are three ways you can work out the volume of the prism above.

1 Count the centimetre cubes (even those you can't see).

2 Work out the volumes of the three separate cuboids that make up the prism, then add them.

3 Work out the area of the cross section and then multiply by the length.

  ×

Use each method and check you get the same answer.

Remember volume is always measured in units^3.

(2) Work out the volume of these shapes. Give your answer to the nearest cm³. (★★★)

a

12 cm

5 cm

(1 mark)

This is a cylinder. It is also a prism. You can find the area of its cross-section (circle) and multiply by the length.

b (2 marks)

15 cm

7 cm

[Total: 3 marks]

SNAP IT! Volume of a cone

volume of a cone $= \frac{1}{3} \times \pi \times$ radius² $\times$ height $= \frac{1}{3}\pi r^2 h$

(3) The volume of a sphere $= \frac{4}{3} \times \pi \times$ radius³ $= \frac{4}{3}\pi r^3$

Work out the volume of this hemisphere.
Give your answer to 2 d.p. (2 marks, ★★★★)

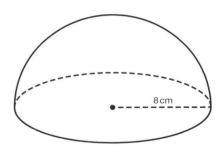

8 cm

(4) The diagram shows a tank in the shape of a cuboid. It also shows a pond in the shape of a cuboid.

The tank is filled so that it is 80% full of water. Then it is tipped into the pond. This is repeated until the pond is full.

How many tanks of water are needed to fill the pond?

Show working to justify your answer. (4 marks, ★★★★★)

60 cm   Tank

40 cm

40 cm

48 cm

Pond

60 cm

80 cm

Surface area

(1) **a** How many faces does a cuboid have? (1 mark, ★)

> The surface area of a shape is the total area of **all** its faces.

....

b What is the surface area of this cuboid? (2 marks, ★★)

....

[Total: 3 marks]

(cuboid labelled 6 cm, 10 cm, 0.5 cm)

(2) The surface area of a pyramid = area of triangular side × 4 + area of square base

Work out the surface area of this square-based pyramid. (2 marks, ★★★)

> Sometimes you may be given more information than you need.
> To find the surface area of this pyramid, you only need two of the lengths that are given.

....

(square-based pyramid labelled 5 cm, 4 cm, 6 cm)

(3) Surface area of a sphere = $4\pi r^2$

Surface area of a cone = area of curved surface + area of circular base

Area of curved surface of a cone = $\pi r \times$ slant length

Work out the surface areas of this sphere and cone. Give your answers to 2 d.p. (★★★)

a (2 marks)

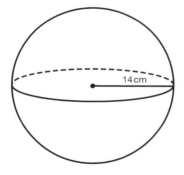

(sphere labelled 14 cm)

b (2 marks)

(cone labelled 8 cm, 10 cm, 12 cm)

....

[Total: 4 marks]

(4) The curved area of a cylinder = $2\pi rh$

Work out the surface area of this shape.
Give your answer to 2 d.p. (2 marks, ★★★★★)

....

> Look at this as two shapes: a cone and a cylinder.
> surface area = curved surface area of cone + curved surface area of cylinder + base of cylinder

(composite shape labelled 11.5 cm, 6 cm, 1.5 cm)

Using Pythagoras' theorem

SNAP IT! Pythagoras' theorem

For a right-angled triangle

$c^2 = a^2 + b^2$

where c is the hypotenuse (the longest side, opposite the right angle) and a and b are the two shorter sides.

(1) Work out the missing side in these two triangles. (2 marks, ★★)

DO IT!

For each triangle.

1 Identify the hypotenuse. Label this c.

2 Substitute the lengths you know into $c^2 = a^2 + b^2$.

3 Don't forget to square root to find the missing side.

....

(2) A rectangular lawn is 4.5 m long. This is 1.5 m shorter than its diagonal.
Work out the area of the rectangle. Give your answer to 1 d.p. (3 marks, ★★★)

> Read the first sentence. Draw a diagram. Read the next sentence. Add the information to your diagram. Underline what the question is asking – the area.

....

(3) Point A has coordinates (2, 1); point B has coordinates (−2, −1). Work out the length of the line AB. Give your answer in surd form. (3 marks, ★★★★)

> Sketch the points on axes. Draw a right angled triangle so that the line AB is the hypotenuse. Now use Pythagoras' theorem. Your answer should be in the form ■√■ .

....

(4) A doorway is 70 cm wide and 190 cm high. A large piece of artwork is 2 m × 2 m. Will it fit through the doorway? You must show your working. (3 marks, ★★★★★)

Trigonometry

(1) *PQR* is a right-angled triangle.

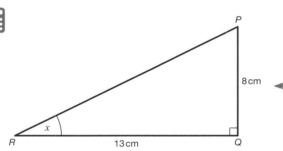

Work out the size of the angle marked x.

Give your answer to 1 d.p. (2 marks, ★★)

Label the opposite, adjacent and hypotenuse sides.

Write

$PQ = 8\,\text{cm} = \ldots\ldots$ (opposite, adjacent or hypotenuse?)

$QR = 13\,\text{cm} = \ldots\ldots$ (opposite, adjacent or hypotenuse?)

Write angle x.

Look at the information you have.

Does the information tell you to use sin, cos or tan?

To find the angle use $\sin^{-1}$ $\cos^{-1}$ $\tan^{-1}$ on your calculator.

....

(2) *ABC* is a right-angled triangle.

Work out the length of side *AC*.

Give your answer to 1 d.p. (3 marks, ★★★)

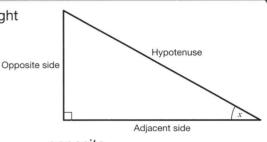

....

(3) A ladder of length 6 m leans against a vertical wall.

The ladder makes an angle of 49° with the ground.

How far is the top of the ladder from the ground?

Give your answer to an appropriate degree of accuracy. (3 marks, ★★★★★)

An appropriate degree of accuracy could be 2 decimal places, as that will give the measurement to the nearest cm.

....

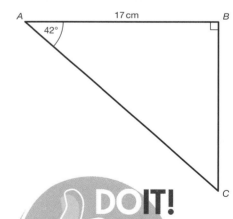

DOIT!

Often it can help to draw a diagram.

1 Draw a diagram of a right-angled triangle.
2 Identify the ladder.
3 Write its length on your diagram.
4 Identify the angle 49°.
5 Write the angle on your diagram.
6 Identify the top of the ladder to the ground.

Exact trigonometric values

SNAP IT! Trigonometric values

You need to learn these trigonometric values:

$\sin 60° = \frac{\sqrt{3}}{2}$ $\cos 60° = \frac{1}{2}$ $\tan 60° = \sqrt{3}$

$\sin 30° = \frac{1}{2}$ $\cos 30° = \frac{\sqrt{3}}{2}$ $\tan 30° = \frac{1}{\sqrt{3}}$

$\sin 45° = \frac{1}{\sqrt{2}}$ $\cos 45° = \frac{1}{\sqrt{2}}$ $\tan 45° = 1$

> Sometimes sine, cosine and tangent are referred to as ratios. The sides in similar triangles are all in the same ratio, producing the same values for sine, cosine and tangent of the angles. Here the ratios are written as fractions.

1. *ABC* is a right-angled triangle. (★★)

   a What is the value of $\tan x$? (1 mark)

   ....

   b What is the size of angle x? (1 mark)

   ....

   [Total: 2 marks]

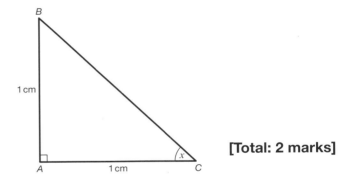

2. *PQR* is a right-angled triangle.

   What is the length of *PR*? (1 mark, ★★★)

   > Label the sides opposite, adjacent and hypotenuse to help you decide whether to use sin, cos or tan.

   ....

3. Triangle *XYZ* has a right angle at *Y*.

   Angle *YXZ* = 30° Side *XZ* = 20 cm

   What is the length of side *YZ*? (2 marks, ★★★★)

   > Draw a diagram of the triangle and put all the information on it.

   ....

4. In a right-angled triangle the ratio between one side and the hypotenuse is $1 : \sqrt{2}$

   What are the sizes of the other angles in the triangle? (2 marks, ★★★★)

   ....

Transformations

(1) On the grid, translate the parallelogram by vector $\begin{pmatrix} -3 \\ -2 \end{pmatrix}$. (2 marks, ★★)

SNAP IT! Translation

A translation is a slide: $\begin{pmatrix} \text{left or right} \\ \text{up or down} \end{pmatrix}$
If the top number is negative, slide left;
if the top number is positive, slide right.
If the bottom number is negative, slide down;
if the bottom number is positive, slide up.

(2) **a** On the grid below right, rotate the triangle 180° clockwise about (−2, 0).

Label the new triangle A. (2 marks, ★★★)

Trace the triangle. Put your pencil on the point (−2, 0) and rotate the tracing paper 180° clockwise.

b On the grid, enlarge your triangle A by scale factor 3 about centre (−2, 1). (2 marks, ★★★)

Label the new triangle B.

Each point on triangle B must be three times as far from point (−2, 1) as the same point on triangle A.

[Total: 4 marks]

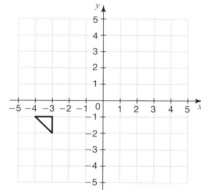

(3) Describe fully the single transformation that maps trapezium A onto trapezium B. (2 marks, ★★★★)

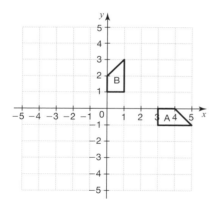

....

....

(4) **a** On the grid, rotate shape P 180° about (1, 1).
Label the new shape Q. (2 marks, ★★★★)

b Reflect shape Q in the line $x = 1$.
Label the new shape R. (2 marks, ★★★★)

Every point on the line $x = 1$ has an x coordinate that is 1.

c Describe fully the single transformation that maps shape R onto shape P. (2 marks, ★★★★)

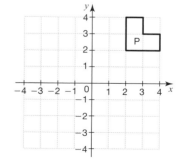

....

[Total: 6 marks]

Similar shapes

SNAPIT! **Similar shapes**

Shapes are similar if one is an enlargement of the other. This means:

- their angles are the same size
- all their lengths are in the same ratio.

(1) Quadrilateral *STUV* is similar to quadrilateral *WXYZ*. (★★)

a In quadrilateral *WXYZ*, which angle is the same size as *UVS* (in quadrilateral *STUV*)? (1 mark)

....

b What is the scale of enlargement that maps quadrilateral *STUV* on to quadrilateral *WXYZ*? (2 marks)

....

scale of enlargement = $\dfrac{\text{enlarged length}}{\text{corresponding original length}}$

c Work out the length of *WZ*. (2 marks)

....

[Total: 5 marks]

(2) Triangle *ABC* is similar to triangle *DEF*. (★★)

Sketch the shapes the same way up. Then you will be able to see the angles that match, and the sides that are in the same ratio.

a Write down the size of angle *EDF*. (2 marks)

....

b Work out the length of *AB*. (2 marks)

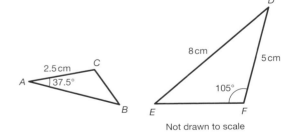

Not drawn to scale

....

c What type of triangles are they? (1 mark)

....

Use what you know about the triangles to find any of the other missing angles or sides. Then you should be able to recognise the type of triangle.

d What is the length of *BC*? (1 mark)

....

[Total: 6 marks]

③ Triangle *PQR* is similar to triangle *SRT*. (★★★★)

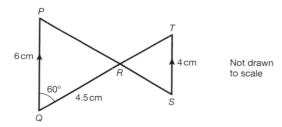

> Sketch the two separate triangles. Make sure you sketch them the same way up.

a What is the scale of enlargement that maps triangle *PQR* on to triangle *SRT*? (2 marks)

....

b Work out the length of *RT*. (2 marks)

....

[Total: 4 marks]

④ The lines *AE* and *BD* are parallel. Triangle *ACE* is similar to triangle *BCD*. (★★★★)

a What is the scale of enlargement that maps triangle *ACE* on to triangle *BCD*? (2 marks)

....

b Work out the length *CE*. (1 mark)

....

c Work out angle *ACE*. (2 marks)

....

[Total: 5 marks]

Vectors

(1) Complete the column vectors. (★★)

$\mathbf{a} = \begin{pmatrix} 3 \\ \square \end{pmatrix}$ (1 mark)　　　$\mathbf{c} = \begin{pmatrix} \square \\ -4 \end{pmatrix}$ (1 mark)

$\mathbf{b} = \begin{pmatrix} -3 \\ \square \end{pmatrix}$ (1 mark)　　　$\mathbf{d} = \begin{pmatrix} \square \\ \square \end{pmatrix}$ (1 mark)

[Total: 4 marks]

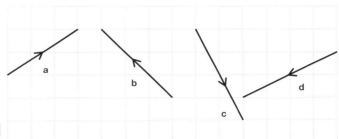

(2) Draw vectors (★★★★)

a **i** $\mathbf{p} = \begin{pmatrix} -6 \\ 1 \end{pmatrix}$　**ii** $-\mathbf{p}$　**iii** $2\mathbf{p}$ (3 marks)

SNAP IT! Magnitudes of vectors

$-\mathbf{p}$ is the negative of $\mathbf{p}$. They have the same magnitude but point in different directions.

$2\mathbf{p}$ is twice the magnitude of $\mathbf{p}$. They have the same direction.

b Write $-\mathbf{p}$ as a column vector. (1 mark)

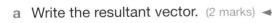

c Write $2\mathbf{p}$ as a column vector. (1 mark)

d Show that $2\mathbf{p} + \mathbf{p} = 3\mathbf{p}$ (3 marks)

Work out $2\mathbf{p} + \mathbf{p} = \begin{pmatrix} -12 \\ 2 \end{pmatrix} + \begin{pmatrix} -6 \\ 1 \end{pmatrix} =$

Work out $3\mathbf{p} = 3 \times \begin{pmatrix} -6 \\ 1 \end{pmatrix} =$

Show that you get the same answer.

[Total: 8 marks]

(3) $\mathbf{a} = \begin{pmatrix} 3 \\ 4 \end{pmatrix}$, $\mathbf{b} = \begin{pmatrix} -5 \\ 1 \end{pmatrix}$, (★★★★)

a Write the resultant vector. (2 marks) ◄ The resultant vector is $\mathbf{a} + \mathbf{b}$

b $\mathbf{c} = \mathbf{a} - \mathbf{b}$ Write $\mathbf{c}$ as a column vector. (2 marks)

c Show $\mathbf{a}$, $-\mathbf{b}$ and $\mathbf{c}$ as a triangle on the grid. (2 marks)

[Total: 6 marks]

Probability
Basic probability

(1) There are ten cookery books on a shelf.

Five are for Indian food.

Four are for Italian food.

One is for Chinese food.

A cookery book is taken at random from the shelf.
What is the probability it is for Chinese food? (1 mark, ★)

> successful outcome = cookery book is for Chinese food
>
> probability = $\dfrac{\text{number of successful outcomes}}{\text{total number of possible outcomes}}$

....

SNAP IT! Probability questions

A probability question may ask for an answer as a fraction, decimal or percentage.

If it doesn't state 'Give your answer as…', and the question includes fractions, then answer with a fraction; if the question includes decimals, then answer with a decimal, and so on.

If it doesn't state 'Give your answer as…' and there is no fraction, decimal or percentage in the question, then you can choose how to write your answer.

(2) The probability it will rain today is 0.6.

What is the probability it will not rain today? (2 marks, ★★)

....

(3) Oliver spins this fair spinner. (★★★)

a On the probability scale, mark A to show the probability that the spinner will land on a number from 1 to 8. (1 mark)

b On the probability scale, mark B to show the probability that the spinner will land on a multiple of 3. (2 marks)

c On the probability scale, mark C to show the probability that the spinner will land on a number greater than 3. (2 marks)

> **NAILIT!**
>
> If the probability of an event occurring is P then the probability of it **not** occurring is $1 - P$.

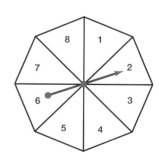

[Total: 5 marks]

(4) The table shows the probabilities of randomly picking different coloured tickets in a lucky dip.

Outcome	Red	Blue	Green
Probability	$2p - 0.1$	$2p + 0.1$	p

Which colour is most likely? You must show your working. (3 marks, ★★★★★)

Write an equation and solve it to find p.

STRETCHIT!

Write and solve an equation where all probabilities add to 1.

Two-way tables and sample space diagrams

1 A factory makes wooden bed frames.

They make singles, doubles and king size.

They use oak, pine or walnut.

The two-way table shows the number of beds the factory makes in one week.

Complete the table. (4 marks, ★★)

	Single	Double	King	Totals
Oak	2			30
Pine		14	17	54
Walnut	1	12		
Totals			32	100

> Look for a row or column with only one number missing. Work out that missing number.

2 A coin is flipped and this fair spinner is spun. (★★★)

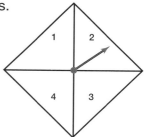

a The sample space diagram shows some of the possible outcomes.

> A sample space diagram is a type of two-way table. It helps identify all possible outcomes for two independent events.
> Events are independent if the outcome of one event does not affect the outcome of another event.

Complete the sample space diagram. (3 marks)

		Spinner			
		1	2	3	4
Coin	**Heads**	1, H			
	Tails			3, T	

b What is the probability of flipping a tail on the coin and spinning a 1 on the spinner? (1 mark)

....

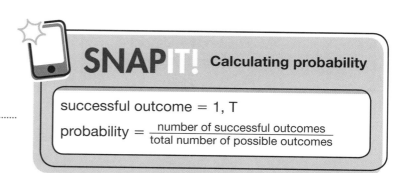

SNAP**IT!** **Calculating probability**

successful outcome = 1, T

probability = $\dfrac{\text{number of successful outcomes}}{\text{total number of possible outcomes}}$

c What is the probability of flipping a head on the coin and spinning a number greater than 1 on the spinner? (1 mark)

....

[Total: 5 marks]

(3) There are 120 students in the sixth form.

$\frac{3}{8}$ of the sixth form are boys.

$\frac{1}{5}$ of the sixth form are boys that study sciences.

$\frac{1}{3}$ of the sixth form are girls that do not study sciences. (★★★★)

a Complete the two-way table. (3 marks)

	Study sciences	Do not study sciences	Totals
Boys			
Girls			
Totals			120

b A student is chosen at random. What is the probability the student is a boy who does not study science? (2 marks)

NAILIT!

When using a fraction to answer a probability question, always cancel the fraction if you can and write it in its simplest terms.

....

c A girl is chosen at random. What is the probability she studies science? (2 marks)

....

[Total: 7 marks]

 STRETCHIT!

When the number of balls in the lottery was increased from 49 to 59, the chances of winning went down to 1 in 45 000 000, or 0.000002%.

The chances of winning are now even lower than the likelihood of being hit by a meteorite!

How could you use probability to help you make sensible decisions?

Sets and Venn diagrams

1. ξ = {integers greater than 20 and less than 30}

   Set A = {multiples of 3 between 20 and 30}

   Set B = {multiples of 4 between 20 and 30} (★★)

   a List all the numbers in the universal set. (1 mark) ◄—— Write ξ = {list all the numbers in curly brackets}

   ....

   b List all the numbers in set A. (1 mark)

   ....

   c List all the numbers in set B. (1 mark)

   ....

   d List all the numbers in A ∪ B. (1 mark)

   ....

   e List all the numbers in A ∩ B. (1 mark)

   ....

   [Total: 5 marks]

2. 20 adults go to the community centre on a Wednesday night and attend either a yoga class, an art class or both.

   8 of them attend a yoga class only.

   5 of them attend the yoga and art class. (★★★)

   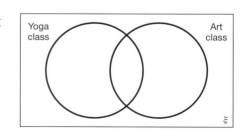

   a Complete the Venn diagram. (3 marks)

   b What is the probability that an adult picked at random goes to the art class only? (2 marks)

   ....

   c What is the probability that an adult picked at random goes to only one class? (2 marks)

   ....

   [Total: 7 marks]

(3) People were surveyed and asked where they bought their food.

S − supermarket

L − local corner shop

F − farmers' market

The Venn diagram shows the responses. (★★★★★)

a One of the people surveyed is picked at random.

Work out the probability they bought their food only from a supermarket. (2 marks)

> First of all, work out how many people were surveyed: ξ = ☐

....

b One of the people surveyed is picked at random.

Work out P(F). (2 marks)

> Don't forget to include everyone in set F.

....

c One of the people surveyed is picked at random. Work out P (L ∩ F). (2 marks)

....

d One of the people surveyed is picked at random.

Work out P(S'). (2 marks)

> S' means everyone **not** in set S (i.e. that do not shop at the supermarket at all).

....

[Total: 8 marks]

Frequency trees and tree diagrams

1. There is one black and three white balls in a bag.

    One ball is taken from the bag, its colour is recorded, ◄───── and then it is put back in the bag.

    A second ball is taken from the bag, and its colour is recorded. (★★★)

    The events are independent, because the first ball is put back in the bag. This means that the outcome of the first ball being taken from the bag does not affect the outcome of the second ball being taken from the bag.

    a Complete the tree diagram. (3 marks)

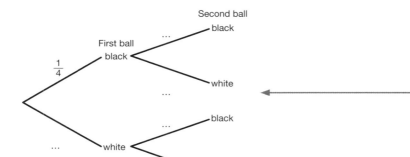

    Write the probability for each event on each branch.

    The probability of the first ball being black has already been written for you.

    b Work out the probability of the first ball being white and the second ball being black. (2 marks)

    NAILIT!

    To use a tree diagram, read along the branches to find each probability and then multiply them together.

    P(W, B) = ...

    c Work out the probability of the first ball being black and the second ball being black. (2 marks)

    P(B, B) = ...

    [Total: 7 marks]

2. There are four apples and two oranges in a fruit bowl. Adam takes a piece of fruit from the bowl and eats it. He then takes another piece of fruit for later. (★★★★)

    a Complete the tree diagram. (3 marks) ◄─────

    The events are **not** independent, because Adam eats the first piece of fruit. This means that the outcome of his first choice affects the possibilities for his second choice.

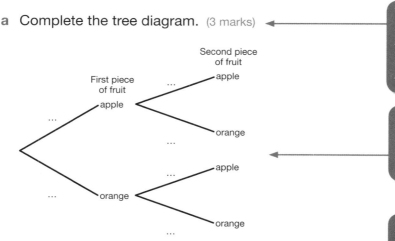

    Write the probability for each event on each branch.

    Be careful ! There are only five pieces of fruit left for Adam's second choice.

    Call the probability of an apple P(A) and the probability of an orange P(O). Add together the probabilities for both possibilities: P (O, O) + P (A, A)

    b What is the probability of the fruit Adam eats being the same as the one he takes for later? (3 marks) ◄

    ....

    [Total: 6 marks]

3 100 students are invited to a careers day.

55 of these students are girls.

4 of the students do not want to go.

42 of the boys want to go. (★★★)

a Use this information to complete the frequency tree. (3 marks)

> You are told how many girls there are, so first work out the total number of boys.

b One of the boys is chosen at random.

Work out the probability that he does not want to go to the careers day. (1 mark)

> Now you can calculate how many boys don't want to go. Anyone else who doesn't want to go must be a girl.

[frequency tree with 100 branching to girls and boys, each branching to "want to go" and "do not want to go"]

[Total: 4 marks]

4 A weather forecaster gives the probability of rain on Saturday as 30% and rain on Sunday as 50%. (★★★★★)

a Draw a tree diagram for the two days, showing the outcomes, 'rain' and 'no rain'. (3 marks)

> Write the percentage probabilities as fractions.

b What is the probability of rain on both days? (2 marks)

c What is the probability of one day having rain and one day having no rain? Give your answer as a percentage. (2 marks)

[Total: 7 marks]

Expected outcomes and experimental probability

(1) The table shows the results of an experiment with this spinner. (★★★)

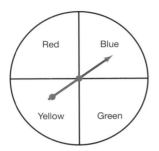

NAILIT!

When writing answers that are fractions, always cancel and give your answer in its simplest form.

Colour	Frequency
Red	12
Blue	13
Green	10
Yellow	15

a How many times in total was the spinner spun? (1 mark)

....

b What is the estimated probability of spinning blue? (2 marks)

....

c What is the estimated probability of spinning yellow? (2 marks)

....

d How many times would you expect to spin green in 100 spins? (2 marks) ◄— P(Green) × 100

....

[Total: 7 marks]

(2) The probability of passing a dance exam is 0.75.

A dance school enters 20 students for the exam. How many should they expect to pass? (1 mark, ★★★)

....

(3) The table shows the reasons customers visit a post office in one hour. (★★★)

Reason	Frequency
Send a parcel	26
Buy stamps	20
Buy foreign currency	6
Banking services	5
Other	3

a Use the data to estimate the probability that in the next hour someone will buy stamps. (2 marks)

....

The post office has, on average, 450 customers each day.

b Estimate the number of people who buy stamps from the post office each day. (1 mark)

....

c Estimate the number of people who buy foreign currency from the post office each day. (2 marks)

....

d Estimate the number of people who use the post office for banking services each week. (The post office is open six days per week). (2 marks)

....

[Total: 7 marks]

Statistics
Data and sampling

(1) A large bakery produces 650 bread rolls each day. The head baker wants to check the quality of a sample of the rolls. How many should she check? Give a reason for your answer. (2 marks, ★)

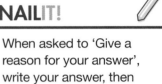

A good sample is 10% of the whole population. The population is the whole group you are interested in. In this case it is all 650 bread rolls.

....

....

(2) A town councillor wishes to investigate how many people use the buses. The population of the town is 25 000. He takes a sample of 50 people in the town centre. Give two reasons why this is not a good sample. (2 marks, ★★★)

....

....

....

NAILIT!

When asked to 'Give a reason for your answer', write your answer, then write 'because…'. Finish the sentence with a reason.

(3) 400 people have accepted an invitation to attend a big street party.

Sam is organising cakes.

She emails a sample of 45 people who are coming to the party.

She asks each one to tell her what type of cake they like.

The table shows information about her results. (★★★★)

Cakes	Number of attendees
Lemon cake	15
Fruit cake	4
Carrot cake	9
Chocolate cake	17

STRETCHIT!

What would Sam do if someone said they didn't eat cake? Consider how she might change the survey to allow for that.

a Work out how many carrot cakes Sam should make. (2 marks)

....

b Write down any assumptions you make and explain how this could affect your answer. (2 marks)

....

....

....

....

....

[Total: 4 marks]

Frequency tables

(1) One evening, a restaurant recorded the number of people sitting at each table. (★★)

a How many tables in the restaurant? (1 mark)

....

b How many people ate at the restaurant that evening? (2 marks)

....

Number of people at a table	Number of tables
1	1
2	11
3	9
4	6
5	1
6	2

[Total: 3 marks]

Look carefully at each column of the table. Understand what each column tells you before answering the questions.

(2) Sarah recorded the number of electronic devices people in her class own. Here are her results.

3, 5, 2, 2, 6, 3, 4, 2, 4, 7, 3, 1, 4, 6, 7, 1, 2, 6, 2, 4, 3, 1, 3, 9

Complete the grouped frequency table with her results. (2 marks, ★★)

As you enter a tally for each number in the table, cross it off the list of results. That way you shouldn't miss any.

Number of electronic devices	Tally	Frequency
0−1		
2−3		

(3) Here are some records of 20 people's weights (in kg) from a database of gym members.

63, 59, 67, 80, 93, 88, 55, 68, 73, 82, 70, 58, 62, 76, 60, 95, 89, 91, 77, 80 (★★★★)

a Is this data discrete or continuous? (1 mark)

....

b Design and complete a grouped frequency table to record the masses. (3 marks)

[Total: 4 marks]

(4) A GP doctor records the length of time she spends with each patient one day.

The GP concludes that 'most appointments are ten minutes long'. Explain why she may be wrong.

(1 mark, ★★★★)

Total number of minutes	Frequency
$0 \leq t < 4$	2
$5 \leq t < 9$	6
$10 \leq t < 14$	12
$15 \leq t < 19$	1

....

Bar charts and pictograms

① The bar chart shows the ways that some university students prefer to exercise. (★★)

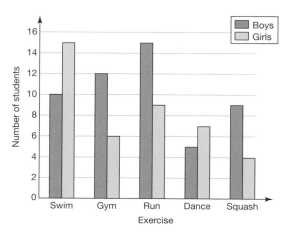

a How many more boys than girls prefer to play squash? (1 mark)

....

b How many girls were surveyed? (2 marks)

....

[Total: 3 marks]

② The table shows the number of suits, trousers and shirts sold each weekend for four weekends in menswear in a department store.

	Suits	Trousers	Shirts
Weekend 1	10	6	6
Weekend 2	7	20	15
Weekend 3	10	10	7
Weekend 4	7	20	12

Complete the composite bar chart. (3 marks, ★★★)

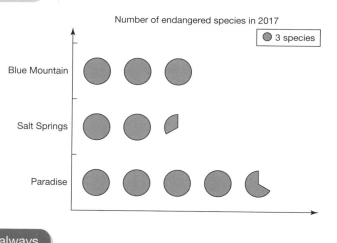

The question asks for a composite bar chart. In this type of bar chart, bars for each category (suits, trousers and shirts) are stacked on top of each other.

③ The pictogram shows numbers of endangered species in three different wildlife reservations in 2017. (★★★)

a How many endangered species are there in Paradise wildlife reservation? (1 mark)

....

b How many more endangered species are there in Blue Mountain than in Salt Springs? (1 mark)

....

For pictograms, always look carefully at the key.

Number of endangered species in 2017

[Total: 2 marks]

Pie charts

(1) 300 people were asked what type of movie they last saw.
The pie chart shows their answers. (★★)

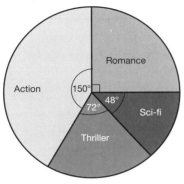

a How many people last saw an action movie? (1 mark)

....

b How many did **not** see a romance movie? (2 marks)

....

[Total: 3 marks]

(2) In one day 250 people visited a castle. The table shows information
about whether they were children, adults or senior citizens.

Children	Adults	Senior citizens
80	155	15

Draw an accurate pie chart for this information. (3 marks, ★★★)

Work out the fraction that were children. That is the fraction
of degrees of the circle you should draw and label children.
(Sometimes it may not be a whole number of degrees.
Then you will need to round to the nearest degree.)

Measure the angle
for salad. How many
degrees is one child?

(3) Some 10-year-old children were surveyed about their favourite
vegetable.

Here is a pie chart of the results.

12 of the children answered salad.

How many children were surveyed? (2 marks, ★★★★)

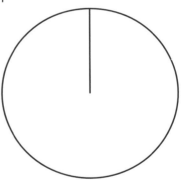

Measures of central tendency: mode

① Here are the times (to the nearest minute) people waited in a queue to buy cinema tickets.

2   4   6   3   3   5   4   4   3   6

What is the modal time? (1 mark, ★)

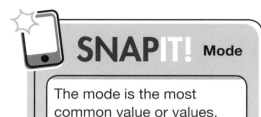

SNAPIT! **Mode**

The mode is the most common value or values.

....

② This table shows the ages of students in a drama club.

What is the modal class? (1 mark, ★★)

Age of students, a (years)	Frequency
$10 < a \leq 11$	5
$11 < a \leq 12$	12
$12 < a \leq 13$	17
$13 < a \leq 14$	10
$14 < a \leq 15$	6

....

③ People were surveyed to find out how many charities they regularly supported.

These are the results.

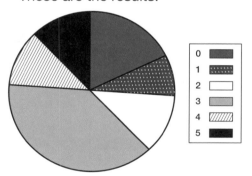

What is the modal number of charities supported? (1 mark, ★★★)

....

④ The bar chart shows the interests that some students have outside school. What is the mode for the boys? What is the mode for the girls? (2 marks, ★)

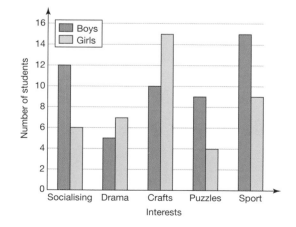

....

Measures of central tendency: median

① Here are the ages of students in a school orchestra.

15 17 13 13 17 11 18 15 12 13 16 18 11 13 16 14

What is the median age? (1 mark, ★★)

....

SNAPIT! Median

The median is the middle value when all values are put in order.

If there are two middle values, then the median lies half way between them.

WORKIT!

A group of people were asked to record how many hours they spent sitting in a day.
The table shows the results.

Time, t (hours)	Frequency
$0 \leq t < 2$	1
$2 \leq t < 4$	1
$4 \leq t < 6$	3
$6 \leq t < 8$	9
$8 \leq t < 10$	13

Which is the median class?

Imagine writing out all the values, in order.

$0 \leq t < 2$ $2 \leq t < 4$ $4 \leq t < 6$ $4 \leq t < 6$ $4 \leq t < 6$ $6 \leq t < 8$...

There would be 1 + 1 + 3 + 9 + 13 = 27 values. (This is the total frequency, or total number of people asked.)

Median (middle) = $\frac{27 + 1}{2}$ = 14th person

The 14th person in the table is in the class $6 \leq t < 8$.

Median class = $6 \leq t < 8$

② This table shows the ages of students in a drama club.

What is the median class? (3 marks, ★★★)

Age of students, a (years)	Frequency
$10 < a \leq 11$	5
$11 < a \leq 12$	12
$12 < a \leq 13$	17
$13 < a \leq 14$	10
$14 < a \leq 15$	6

....

Measures of central tendency: mean

(1) Here are the ages of children in a family.

6   7   11   13   18

What is their mean age? (1 mark, ★★)

SNAPIT! **Mean**

$$\text{mean} = \frac{\text{sum of all values}}{\text{number of values}}$$

....

(2) The bar chart shows the number of bedrooms in houses on a newly built estate.

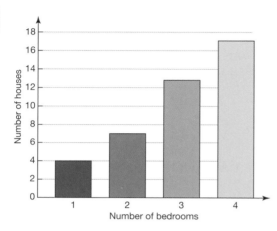

Number of houses (y-axis)
Number of bedrooms (x-axis)

Work out the mean number of bedrooms.

(3 marks, ★★★)

How many houses have 1 bedroom? How many bedrooms in total do they have?

How many houses have 2 bedrooms? How many bedrooms in total do they have?

Repeat these calculations for each number of bedrooms to work out the sum of all the bedrooms.

Divide by the total number of houses.

....

STRETCHIT!

What would you like to find out about? It's a serious question, because the whole point of maths is to work out something you want to know.

Examples in books are written to help you learn the skills. In real life, you write your own questions – and are genuinely interested in the answers!

③ The frequency table shows the number of holidays a company's employees have taken in a year.

Work out the mean number of holidays for all employees. (3 marks, ★★★)

Number of holidays	Frequency
0	4
1	21
2	9
3	2

NAIL IT!

Think about the information you are working with. Does it make sense to give an answer that is a decimal or a fraction? If not, then round to the nearest whole number.

How many people took 0 holidays? How many holidays in total is that?

How many people took 1 holiday? How many holidays in total is that?

Put an extra column on the frequency table to help you work this out: **number of holidays × frequency**

Work out the sum of all the holidays.

Divide by the total number of employees (total frequency).

....

④ The grouped frequency table shows the ages of patients waiting to be seen in a hospital accident and emergency department.

What is the mean age of the patients? (2 marks, ★★★★)

Create two new columns to help you work out the answer.

Age of patients (a)   **Midpoint**   **Frequency**   **Midpoint × Frequency**

Age of patients (a)	Frequency
$0 < a \leq 10$	3
$10 < a \leq 20$	18
$20 < a \leq 30$	6
$30 < a \leq 40$	11
$40 < a \leq 50$	10
$50 < a \leq 60$	19
$60 < a \leq 70$	16
$70 < a \leq 80$	17

....

Range

① The bar chart shows the ages of children in a nursery. (★)

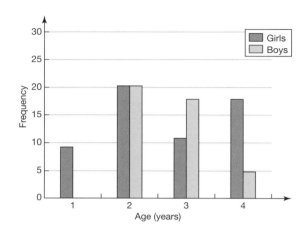

a What is the range of ages of the boys? (1 mark)

....

b What is the range of ages of the girls? (1 mark)

....

[Total: 2 marks]

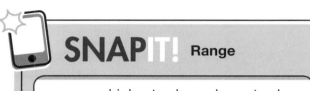

SNAPIT! **Range**

range = highest value − lowest value
Don't forget to include units when giving your answer. For example, year, °C, etc.

② The table shows average midday temperatures for ten weeks during the summer. (★★)

Week	1	2	3	4	5	6	7	8	9	10
Temperature °C (Resort A)	17	17	18	21	25	20	21	18	16	16
Temperature °C (Resort B)	13	15	18	23	27	27	28	22	16	14

a What is the range in temperatures for Resort A? (1 mark)

....

b What is the range in temperatures for Resort B? (1 mark)

....

[Total: 2 marks]

③ Here are the profit figures for two small businesses. (★★)

	Year 1	Year 2	Year 3	Year 4
Business A	£23 561	£30 485	£39 210	£45 816
Business B	£32 820	£40 328	£17 894	£63 248

a What is the range and mean yearly profit for Business A? (2 marks)

....

b What is the range and mean yearly profit for Business B? (2 marks)

....

There are 2 marks for part c, so give two reasons for your choice. Use the information you have found in parts a and b.

c Which business is doing better? Give reasons for your answer. (2 marks)

....

....

[Total: 6 marks]

Comparing data using measures of central tendency and range

(1) The time taken for the same bus journey each morning, Monday to Saturday, is recorded in the table below.

Monday	Tuesday	Wednesday	Thursday	Friday	Saturday
32 minutes	30 minutes	39 minutes	32 minutes	43 minutes	31 minutes

The same journey can be made by train. The time taken for this train journey, Monday to Saturday, is recorded in the table below. (★★)

Monday	Tuesday	Wednesday	Thursday	Friday	Saturday
16 minutes	24 minutes	18 minutes	26 minutes	70 minutes	17 minutes

a Work out the mean time for the bus journey. (2 marks)

b Work out the mean time for the train journey. (2 marks)

c Which journey has the bigger range? (1 mark)

d Is it better to get the bus or the train? Give reasons for your answer. (1 mark)

[Total: 6 marks]

(2) Two council officers record the ages of people at a playground as

5 4 33 4 8 26 3 7 30 7 8 5 3 39 (★★★★)

NAILIT!

When asked to 'explain', use calculations to support your explanation.

a One council officer finds the mean to report the average age of playground users. Explain why this is not the best average to use. (2 marks)

b The other council officer says that they should use the mode to report the average age of playground users. Explain why this is not the best average to use. (2 marks)

c The council officers should use the median. Work out the median. (1 mark)

[Total: 5 marks]

(3) There are 20 students in a class. The class teacher records the number of days each student is absent due to sickness in one term:

0 0 0 1 0 1 2 0 5 0 0 1 0 24 2 1 0 0 0 3

Which average should the class teacher use to describe absences due to sickness in her class?

Explain why she should use this average. (3 marks, ★★★★)

....

....

....

NAILIT!

Look at the number of marks a question is worth. If it is worth
4 marks, then try to include four calculations and/or statements.

Time series graphs

① The time series graph shows the weight, in grams, of a male kitten for the first seven days of his life. (★★★)

Weight (grams) per day of a male kitten

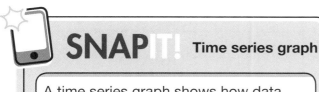

SNAPIT! **Time series graph**

A time series graph shows how data changes over time. Time is usually plotted on the horizontal (x) axis.

a What did the kitten weigh on day 3? (1 mark)

....

b How old was the kitten when he weighed 220 g? (1 mark)

....

mean weight = total weight gained / number of days

c What was the mean weight the kitten gained each day in the first week of his life? (2 marks)

....

[**Total: 4 marks**]

② A campsite opens for the season in March. The table shows the number of campers on the campsite for the first six months. (★★★)

Month	Mar	Apr	May	Jun	Jul	Aug
Number of campers	30	62	125	80	108	150

a Draw a time series graph to represent the data. (3 marks)

 NAILIT!

1 Write a title for your graph.

2 Choose a sensible scale for the number of campers (going up in steps of two will make it a very big graph; going up in steps of 100 will make it a very small graph).

3 Label your axes.

4 Plot your points carefully.

5 Join your points with straight lines.

b What do you notice about May? Give two different sensible reasons for this happening. (2 marks)

....

....

[**Total: 5 marks**]

③ Two new branches of a DIY store open in two different towns. The table shows the sales figures for each new DIY store in its first eight months. (★★★★)

Month	1	2	3	4	5	6	7	8
Branch 1 sales (£)	£6450	£11740	£16690	£25870	£24920	£25560	£26100	£25930
Branch 2 sales (£)	£4390	£6210	£5950	£20380	£23820	£27460	£32460	£37590

a Draw a time series graph for each shop's sales on the same axes. (3 marks)

b Describe the trend in sales for each branch for the eight months. (4 marks)

> There are 4 marks for part **b**, so you need to write at least two things about Branch 1; and at least two things about Branch 2.

....

....

....

....

[Total: 7 marks]

Scatter graphs

(1) Here are two scatter diagrams.

Scatter diagram A shows temperature (°C) and pairs of flip flops sold.

Scatter diagram B shows temperature (°C) and pairs of wellington boots sold. (★★)

 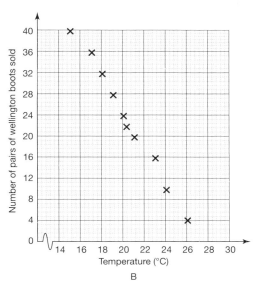

a Describe the correlation for scatter diagram A. Give a reason for your answer. (2 marks)

....

b Describe the correlation for scatter diagram B. Give a reason for your answer. (2 marks)

....

[Total: 4 marks]

(2) The table shows the percentage achieved by the same students in a
maths test and a physics test. (★★★)

Maths test (%)	45	37	55	70	0	62	48	59	89	30	55	43
Physics test (%)	40	43	64	79	58	66	52	65	80	42	60	40

a Draw a scatter diagram for this data. (3 marks)

b Describe what your scatter diagram shows. (2 marks)

....

....

....

State whether the correlation is positive or negative. As it is worth 2 marks, describe what this means, i.e. what can you say about how the students who scored higher marks in the maths test did in the physics test?

c Circle the outlier. (1 mark)

d Give a sensible reason for the outlier. (1 mark)

....

[Total: 7 marks]

(3) A second-hand computer shop recorded laptops' selling prices against their age.

The scatter diagram shows the results. (★★★★)

Age of laptop (months)

NAILIT!

When drawing a line of best fit, try to ensure the same number of crosses is on each side of the line. There may or may not be crosses on the line too.

a The shop owner tells an employee that, on average, a laptop loses £150 every 6 months.

Draw a line of best fit on the scatter diagram. Does it agree with what the shop owner says?

Is he correct? Show how you get your answer. (3 marks)

....

....

....

....

....

....

b You should not use a line of best fit to predict the price of a laptop at 20 months. Give one reason why. (3 marks)

....

....

[Total: 6 marks]

Use your own paper to answer these questions.

Foundation tier

Time: 1 hour 30 minutes

The total mark for this paper is 80.
The marks for **each** question are shown in brackets.

1   Circle the answer to the calculation 0.34×10.

    0.34   0.340   3.4   34   **[1 mark]**

2   Which of the shapes B – E is congruent to shape A?

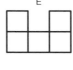

    [1 mark]

3   Circle the fraction that is **not** equivalent to $\frac{4}{7}$.

    $\frac{10}{14}$ $\frac{12}{21}$ $\frac{16}{28}$ $\frac{24}{42}$

    [1 mark]

4   Simplify $4x - (2 - 3x)$.

    Circle your answer.

    $2 - x$ $7x - 2$ $-x$ $x - 2$ **[1 mark]**

5   Here is a calculation: $\frac{7.93 + 12.1}{3.86}$.

    Circle an estimate for the answer to this calculation.

    14   10   5   4   **[1 mark]**

6   Sara says, 'These two fractions, this decimal and this percentage are all equivalent.'

    Sara is incorrect.

    Circle the one **not** equivalent to the others.

    $\frac{4}{5}$ $\frac{8a}{10a}$ 8%   **[1 mark]**

7   Tick two facts about this quadrilateral:

☐ It is a rectangle.

☐ Its opposite sides are equal.

☐ Its diagonals bisect at right angles.

☐ It is a rhombus.

☐ It is a parallelogram.   [2 marks]

8   Work out $\frac{3}{8}$ of 2400.   [2 marks]

9   The ratio of girls to boys in a class is 5:7.

     Write the fraction of the class that are boys.   [1 mark]

10   Here are some numbers:

     24   12   9   20   18

     a Circle the numbers that are factors of 36 **and** multiples of 6.   [2 marks]

     b Write **all** the prime factors of 36.   [2 marks]

     c Is it possible to write a prime multiple of 6?

                 ☐ Yes   ☐ No

     Give a reason for your answer.   [2 marks]

                                                                   [Total: 6 marks]

11   Kamal has five coins.

     Two of the coins are the same.

     Four of the coins add up to 70p.

     All of the coins add up to £1.70.

     What coins does Kamal have?   [3 marks]

12   Write 25 minutes as a fraction of 1 hour.

     Give your answer in its simplest form.   [2 marks]

13   Here are three expressions:

$a + 2b$ $2(a - b)$ $\dfrac{2a}{b}$

$a = 6$ and $b = 4$

Which expression gives an odd number answer?

You **must** show your working.   [2 marks]

14   The perimeter of this triangle *ABC* is three times as many units as the area of this rectangle *PQRS*.

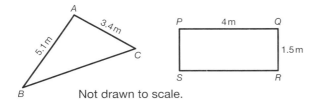

Not drawn to scale.

What is the length of *BC*?   [4 marks]

15   In a pizza restaurant, the ratio of adults to children is $2:3$.

a   There are 16 adults. How many children are there?   [2 marks]

b   Another 4 adults decide to eat at the restaurant.

What is the ratio of adults to children now?

Give the ratio in its simplest form.   [2 marks]

[Total: 4 marks]

16   A tin is 20 cm × 4 cm × 1 cm.

Some colouring crayons are 10 cm × 8 mm × 1 cm.

How many colouring crayons can you fit in the tin?   [2 marks]

17   On a farm, 10% of the animals are horses; 12% are cows; 18% are chickens.

The rest are sheep. There are 240 sheep.

How many animals are on the farm?   [3 marks]

18 In a café, Antony buys two coffees and a cake for £4.40.

Naomi pays Antony £3.20 for one coffee and the cake.

How much was the cake?   **[3 marks]**

19 Here is the graph of $y = 3x + 1$ for values of x from -1 to 1.

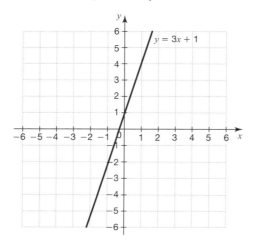

a On the same grid, draw the graph of $y = x + 2$ for values
   of x from -2 to 2.   **[2 marks]**

b Use the graphs to solve the simultaneous equations

$y = 3x + 1$ and $y = x + 2$.   **[2 marks]**

                                                          [Total: 4 marks]

20 A taxi fare is £3.50 flat fee and then £2.25 per kilometre.

Abbie pays £12.50 for a taxi ride.

How far did Abbie travel in the taxi?   **[3 marks]**

21 A marketing company asks 90 people how best to contact them.

The table shows the results.

Phone call	Letter	Text	Email
10	5	45	30

Ali uses this information to draw a pie chart of the results.

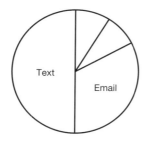

Write down two different mistakes that Ali has made.   **[2 marks]**

22   a Write down all whole number values that satisfy the following inequalities:

        i $-2 \leq a < 3$ ii $-5 < b \leq 2$ **[2 marks]**

    b Work out the largest possible whole number value for ab.   **[2 marks]**

                                                                       [Total: 4 marks]

23   Work out 1.5×0.0003.

    Write your answer in standard form.   **[2 marks]**

24   The bar chart shows the percentage of fiction books, local history books and other factual books for loan in two different local libraries.

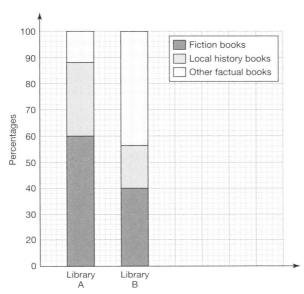

Library A has 1500 books in total.

Library B has 3200 books in total.

    a A book is picked at random from Library A.

        What is the probability it is **not** a fiction book?   **[2 marks]**

    b Which library has more local history books?

        You **must** show your working.   **[3 marks]**

                                                           [Total: 5 marks]

25   Work out $3\frac{3}{4} + 1\frac{7}{8}$.   **[3 marks]**

26   a Solve $3(x + 2) = 5x - 4$.   **[2 marks]**

    b Factorise $3x^2 + 27x$.   **[2 marks]**

                                                           [Total: 4 marks]

27   Work out the area of this triangle.   **[4 marks]**

6 cm

10 cm

28   The nth term of a sequence is $5n + 1$.

    a   What is the position of the term 36 in the sequence?   [1 mark]

    b   Circle the sequence that also includes the term 36:

        $3n - 1$ $2n + 3$ $5n - 8$ $7n + 1$

    You **must** show your working.   [2 marks]

29   Here are sketches for three triangles.

In each triangle, one length is 10 cm.

The other length is given to 3 significant figures.

    a   Circle the value of tan 55° to 3 significant figures.

        0.819   8.19   1.43   5.74   [1 mark]

    b   Circle the value of sin 55° to 3 significant figures.

        0.819   8.19   1.43   5.74   [1 mark]

                                                      [Total: 2 marks]

30   This shape is a semi-circle with diameter 8 cm, joined to a trapezium.

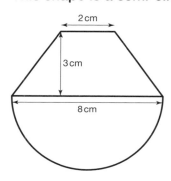

Work out the area of the shape.

Give your answer in terms of π.   [4 marks]

                                                     [Total marks: 80]

Practice paper (calculator)

Use your own paper to answer these questions.

Foundation tier

Time: 1 hour 30 minutes

The total mark for this paper is 80.
The marks for **each** question are shown in brackets.

1   Which of these is a prime number?

       Circle your answer.

       1   8   11   15   **[1 mark]**

2   Which two triangles are similar?

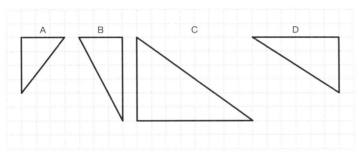

                                                           [1 mark]

3   Which number is 7 more than −10.2?

       −17.2   −3.8   3.8   −3.2   **[1 mark]**

4   Which is the better rate of pay for a 37 hour week?

       A £11.50 per hour   or   B £462.50 per week.

       You **must** show your working.   **[2 marks]**

5   A bus takes 21 minutes and 25 seconds from the bus station to the hospital.

       The bus arrives at the hospital at 2.10pm.

       Exactly what time did the bus leave the bus station?   **[1 mark]**

6   The points (2, 3), (−4, 2) and (−2, −2) are three vertices of a parallelogram.

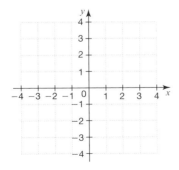

       Work out the coordinates of the other vertex.   **[2 marks]**

7   In a quiz, teams score out of 20.

Here are the scores from six teams.

18   16   19   2   18   17

a Work out the mean of the six scores.   [1 mark]

b What is the median of the six scores?   [1 mark]

c Explain which is the better average for this data: mean or median?   [2 marks]

[Total: 4 marks]

8   The weather forecast says the probability of snow is 20%.

What is the probability that is **does not** snow?

Give your answer as a decimal.   [2 marks]

9   The angles in a triangle are in the ratio 2:2:5.

a Sketch the triangle and write the angles on your sketch.   [2 marks]

b What type of triangle is this?   [1 mark]

[Total: 3 marks]

10   A gardener buys 4 m of string.

He uses $\frac{3}{8}$ of the string to tie some plants.

He uses $\frac{4}{5}$ of the remaining string to lay out the edges of a flower bed.

How much string does the gardener have left?

Give your answer in centimetres.   [3 marks]

11   The bar chart shows the annual sales and costs (in £ thousands) for three different
branches of the same shop.

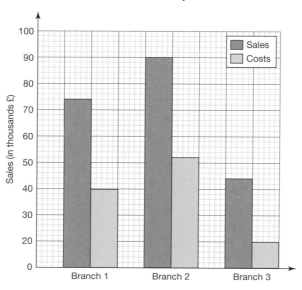

a Work out the profit for Branch 1.   [1 mark]

b The manager of Branch 2 says,

'My sales are over twice what Branch 3 have sold. This means I have made double the profit of Branch 3.'

Is the manager correct?

You **must** show your working.   [2 marks]

[Total: 3 marks]

12 a Solve $\frac{x}{5} = 20$.   [1 mark]

b Simplify $3 \times a \times a + 2 \times a \times b$.   [1 mark]

c A box contains t tins of beans and costs p pounds.

Write a formula for c, the cost of one tin of beans.   [2 marks]

[Total: 4 marks]

13 A plumber charges a call out fee of £70.

He then charges £35 per hour.

He adds 20% VAT to his total bill.

Amy calls out the plumber. He takes 3 hours to do the work.

How much is Amy's plumbing bill?   [3 marks]

14 Roman, Freddie and Claudia are in the same family.

Their ages add up to 33.

The total of Roman's and Freddie's ages is 27.

Roman is nine years older than Claudia.

Work out Freddie's age.   [3 marks]

15 Here is a quadrilateral:

Angle y is twice the size of angle x.

a Work out the size of angles x and y.   [3 marks]

b What type of quadrilateral is this?   [1 mark]

[Total: 4 marks]

16   A parking meter takes 10p, 20p and 50p coins.

In the meter there are four times as many 50p coins as 20p coins.

There are half as many 10p coins as 20p coins.

There are eight 10p coins in the meter.

How much is in the meter?   **[3 marks]**

17   Jeff hires deckchairs in the park.

He draws this scatter diagram of the temperature against the number of deckchairs hired each day for two weeks.

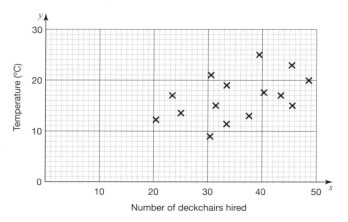

a What type of correlation does the diagram show?

Circle your answer.

Positive correlation   Negative correlation   No correlation   **[1 mark]**

b If the temperature is 20°C, how many deckchairs can Jeff expect to hire?

You must show your working.   **[3 marks]**

                                                                        [Total: 4 marks]

18   The Venn diagram shows the numbers of students studying languages in Year 11 in a school.

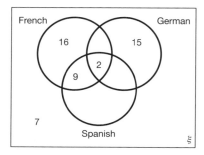

a 33 students study French

28 students study German

40 students study Spanish

Complete the Venn diagram.   **[3 marks]**

b How many students are there in total?   [1 mark]

c A student is chosen at random. Write down P(German ∩ Spanish).   [2 marks]

[Total: 6 marks]

19 In a sale, the original price of a bed is reduced by $\frac{2}{9}$.

The sale price is £350.

Work out the original price.   [2 marks]

20 A large sculpture is made of two bronze spheres.

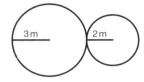

The left-hand sphere has a radius of 3 m.

The right-hand sphere has a radius of 2 m.

The volume of a sphere is $\frac{4}{3}\pi r^3$.

Work out the volume of bronze used to make the sculpture.

Give your answer to 2 decimal places.   [4 marks]

21 $\mathbf{a} = \begin{pmatrix} -3 \\ 2 \end{pmatrix}$ $\mathbf{b} = \begin{pmatrix} -1 \\ -4 \end{pmatrix}$

Circle the vector $\mathbf{a} - \mathbf{b}$.

$\begin{pmatrix} -2 \\ -2 \end{pmatrix}$ $\begin{pmatrix} 2 \\ -6 \end{pmatrix}$ $\begin{pmatrix} -4 \\ -6 \end{pmatrix}$ $\begin{pmatrix} -4 \\ -2 \end{pmatrix}$ [1 mark]

22 Lucy buys 500 US dollars ($) for her holiday.

She gets an exchange rate of £1 = $1.272.

a How much in GB pounds (£) does it cost Lucy to buy $500?   [2 marks]

Lucy spends 427 US dollars on holiday.

b Lucy brings the rest of the dollars home.

She converts them back to GB pounds. For this, she gets a rate of $1 = £0.789.

How much money (in £) does Lucy get when she returns home?   [3 marks]

[Total: 5 marks]

23 XYZ is a right-angled triangle.

YZ is 7 cm.

Angle XZY is 42°.

Work out the hypotenuse of the triangle.

Give your answer to 2 decimal places.   [3 marks]

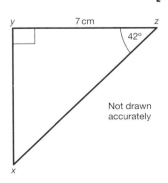

361

24   x^3 equals a negative number.

Tick a box.

☐ x is a negative number   ☐ x is a positive number

Reason _____   **[1 mark]**

25   a A line of trees is planted so that it is the same distance from a castle, C, and the
      gatehouse, G. Using ruler and compasses, draw the line of trees.
                                                                          [3 marks]

C
×

G
×

b The scale for the diagram is 1 cm represents 200 m. Work out the distance
between the castle and the gatehouse. Give your answer in km.   **[2 marks]**

                                                                  [Total: 5 marks]

26   a Use your calculator to work out $4.927^2 + \sqrt[3]{8.135}$. Give your answer
      to 3 decimal places.   **[1 mark]**

b Use approximation to check that your answer to part a is sensible.

You **must** show your working.   **[2 marks]**

                                                                  [Total: 3 marks]

27   The value of a car is £17 000.

The value depreciates by 18% per year.

Work out the value of the car after 3 years.

Give your answer to the nearest pound. **[3 marks]**

28   Here is a straight line, AB

a Write the equation of line AB.   **[2 marks]**

b Which of these lines is parallel to AB?

Circle your answer.

$y = x + 2$ $y = 2x + 1$
$y = -2x$ $y = 2x$ **[1 mark]**

                                                                  [Total: 3 marks]

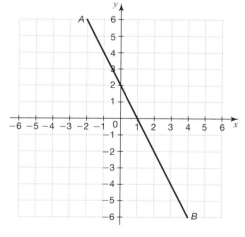

                                                                  [Total marks: 80]

Answers

For full worked solutions, visit:
www.scholastic.co.uk/gcse

Number

Factors, multiples and primes

1 6
2 17, 19, 23
3 $60 = 2^2 \times 3 \times 5$
4 Drummer 1 hits her drum at: 6 12 18 24 30 36 42 48 54 60 seconds
   Drummer 2 hits his drum at: 8 16 24 32 40 48 56 seconds
   They hit their drums at the same time twice (two times) after 24 and 48 seconds.

Ordering integers and decimals

1 $-12, -8, -1, 0, 2$
2 0.32, 0.3, 0.23, 0.203
3 **a** $-4 < 0.4$ **c** $-0.404 > -0.44$
   b $4.200 < 4.3$ **d** $0.33 < 0.4$

Calculating with negative numbers

1 **a** -10 **b** -4 **c** 5   **d** 1
2 **a** -18 **b** 4   **c** 40   **d** -16
3 22
4 1 correct answer; 4 incorrect answers

Multiplication and division

1 **a** 2142   **b** 11 223   **c** 92   **d** 52
2 **a** 12   **b** 12
3 £335
4 1196 hours

Calculating with decimals

1 76.36
2 £7.51
3 38.29
4 Flo raises £28.75; Kirsty raises £143.75

Rounding and estimation

1 **a** 0.798   **b** 0.80
2 5
3 **a** £7500
   b Overestimate, because the concert ticket price and number of tickets sold were rounded up, and so the amount of income was estimated more than it really is.

Converting between fractions, decimals and percentages

1 **a** $\frac{71}{1000}$ **c** 40%
   b 0.63   **d** $\frac{8}{25}$
2 **a** 0.3125   **b** 31.25%
3 $\frac{5}{8} = 0.625$ 0.65 $60\% = 0.6$
   Therefore, 0.65 is largest.

Ordering fractions, decimals and percentages

1 **a** $\frac{1}{2} < 0.6$ **b** $\frac{3}{4} > 0.7$ **c** $-\frac{3}{10} < 0.2$
2 **a** $\frac{5}{12}$ $\frac{9}{20}$ $\frac{7}{15}$
   b $\frac{1}{25}$ 0.4   45%
3 $\frac{1}{3} = 33.3\%$; $\frac{2}{5} = 40\%$, so shop C, shop A, shop B
4 $\frac{5}{9}$ 38.5% 0.38 $\frac{3}{10}$

Calculating with fractions

1 $\frac{29}{45}$ 2 $\frac{1}{12}$ 3 $\frac{11}{21}$ 4 10

Percentages

1 10
2 £13.60
3 14 193
4 £1008

Order of operations

1 7   2 23   3 4.0964

Exact solutions

1 $0.133\,\text{cm}^2$
2 $1\frac{7}{9}\,\text{m}^2$
3 $2\sqrt{3}\,\text{cm}^2$
4 Area of a circle $= \pi r^2$
   The fraction of the circle shown $= \frac{3}{4}$
   The area of the circle shown $= \frac{3}{4} \times \pi r^2$
   Radius = 2 cm
   The area of the shape $= \frac{3}{4} \times \pi \times 2^2 = \frac{3}{4} \times \pi \times 4 = 3\pi$

Indices and roots

1 **a** 7^4 **b** 5^{-3}
2 **a** 16   **b** $\frac{1}{100}$
3 $3^{-2} = \frac{1}{9}$ $\sqrt[3]{27} = 3$ $\sqrt{25} = 5$ $2^3 = 8$
4 1

Standard form

1 2750   3 6.42×10^{-3}
2 1.5×10^8 4 $2.8 \times 10^{-4}\,\text{km}$

Listing strategies

1 259, 295, 529, 592, 925, 952
2 **a**

		4-sided spinner			
		0	1	2	3
3-sided spinner	1	1	2	3	4
	2	2	3	4	5
	3	3	4	5	6

   b 3
3 15
4 spj; spi; sfj; sfi; bpj; bpi; bfj; bfi

Algebra

Understanding expressions, equations, formulae and identities

1 **a** identity   **b** equation   **c** expression
2 **a** equation, because it has an equals sign and can be solved
   b formula, because it has letter terms, an equals sign and the values of the letters can vary
   c an expression because it has letter terms and no equals sign
   d formula, because it has letter terms, an equals sign and the values of the letters can vary
3 **a** Any of: $2x + 10$ or $10x + 2$ or $x + 210$ or $x + 102$
   b Any of: $2x = 10$ or $10x = 2$

Simplifying expressions

1 $8x$
2 **a** $48a^2$ **b** $30p^3$
3 $5y$
4 $8u$

Collecting like terms

1. **a** $3m + 4n$ **b** $-3q - 2r$
2. **a** $a + 13b$ **b** $-c + d$
3. **a** $5p^3 + p$ **b** $-7x^2 + 3x + 12$ **c** $3m + n^2$
4. $-5\sqrt{5} + f$

Using indices

1. **a** p^4 **b** $12y^5$ **c** $10a^5b^3$
2. **a** q^{-6} **b** u^{-6} **c** 1
3. **a** b **b** f^3 **c** $\frac{y^2}{x}$
4. $(2m^3)^3$

Expanding brackets

1. **a** $4m + 12$ **b** $2p - 2$ **c** $30x - 50$
2. **a** $8m + 11$ **b** $4x + 2$
3. **a** $y^2 + 10y + 21$
   b $b^2 - 2b - 8$
   c $x^2 - 10x + 24$
4. **a** $q^2 + 2q + 1$ **b** $c^2 - 6c + 9$ **c** $12m^2 + mn - n^2$

Factorising

1. **a** $4(x + 2)$ **b** $3(d - 5)$ **c** $4(2y - 3)$
2. **a** $q(q + 1)$ **b** $a(a + 6)$ **c** $5z(2z + 3)$
3. **a** $(x + 3)(x + 4)$
   b $(x - 2)(x + 8)$
   c $(a - 6)(a - 4)$
4. **a** $(y + 2)(y - 2)$
   b $(x + 3)(x - 3)$
   c $(p + 10)(p - 10)$

Substituting into expressions

1. 2
2. 44
3. **a** -3 **b** -28 **c** 8 or -8
4. -3

Writing expressions

1. **a** $n + 3$ **b** $2n - 9$
2. **a** $x + y$ **b** $5x$ **c** $12x + 11y$
3. $28p + 4$
4. $s(5s + 1)$

Solving linear equations

1. **a** $x = 7$ **b** $x = 13$ **c** $x = 5$ **d** $x = 18$
2. **a** $x = 6$ **b** $x = 7$ **c** $x = 25$ **d** $x = 3$
3. **a** $x = 1$ **b** $p = 4$ **c** $m = 2$ **d** $q = 4$
4. **a** $x = 5$ **b** $y = 4$ **c** $x = 3$ **d** $n = 6$

Writing linear equations

1. $x = 20$
2. 12 years old
3. $130\,\text{cm}^2$
4. $x = 15, y = 4$

Linear inequalities

1. **a** $-1, 0, 1, 2, 3, 4, 5$
   b
   $-3\ -2\ -1\ \ 0\ \ 1\ \ 2\ \ 3\ \ 4\ \ 5\ \ 6$
2. **a** $x > 5$
   $4\ \ 5\ \ 6\ \ 7\ \ 8\ \ 9\ \ 10$
   b $x \le 7$
   $4\ \ 5\ \ 6\ \ 7\ \ 8\ \ 9\ \ 10$
   c $x < 8$
   $4\ \ 5\ \ 6\ \ 7\ \ 8\ \ 9\ \ 10$
3. **a** $-1 \le x < 2$
   b $-2 < x \le 0$
4. $1, 2, 3, 4, 5$

Formulae

1. **a** 1 hour 40 minutes **b** 12.10 pm
2. **a** $C = l + nk$ **b** £109.50
3. **a** $q = \frac{3p}{s}$ **c** $q = \frac{p - 3r}{3}$
   b $q = rp - rt$ **d** $q = \frac{p^2}{2}$

Linear sequences

1. **a** 6, 11, 16, 21 **b** 251
2. **a** 19
   b No, Rachel is not correct, because the number of triangles is not the pattern number multiplied by 2. Instead, it is the pattern number add 2, so there will be 6 triangles in pattern 4.
3. **a** $11n - 8$
   b No, because then $11n - 8 = 100$ so $11n = 108$, n is not a whole number.

Non-linear sequences

1. **a** 16, 32 **b** 0.1, 0.01 **c** 24, -48 **d** geometric
2. **a** 15 **b** 9th
3. **a** Yes, he is correct.

Day	Mon	Tue	Wed	Thu	Fri
Number of ladybirds	2	8	32	128	512

   b Saturday
4. **a** $\frac{1}{2}, 2, 4\frac{1}{2}$
   b Yes, because $\frac{1}{2}n^2 = 32$, $n^2 = 64$, $n = 8$

Show that...

1. LHS $= 2x + 1$; RHS $= 2x + 1$; LHS = RHS. Therefore, $2\left(x + \frac{1}{2}\right) \equiv x + x + 1$
2. LHS $= x^2 - 25 + 9 = x^2 - 16$; RHS $= x^2 - 16$
3. Let the three consecutive numbers be n, $n + 1$ and $n + 2$. $n + n + 1 + n + 2 = 3n + 3 = 3(n+1)$. Therefore, the sum of three consecutive numbers is a multiple of 3.
4. **a** width of pond $= x - y + x + x + x - y = 4x - 2y$
   length of pond $= 4x$
   Perimeter $= 4x - 2y + 4x - 2y + 4x + 4x = 16x - 4y$
   b Yes Sanjit is correct, because $16x - 4y = 4(4x - y)$, showing that when x and y are whole numbers, the perimeter is always a multiple of 4.

Functions

1. **a** 11 **b** 6 **c** $y = 4x - 1$

2.

x	y
-2	-1
0	3
3	9

3.

x	y
-2	0
1	$1\frac{1}{2}$
8	5

Coordinates and midpoints

1. **a** $(1, 2)$
   b

   c $\left(4\frac{1}{2}, 1\right)$

2 a (1, −4)

b

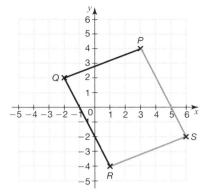

3 a (1, 3) **b** $\left(4, \frac{1}{2}\right)$ **c** $\left(1, -1\frac{1}{2}\right)$

Straight-line graphs

1 a

x	−1	0	1	2
y	1	3	5	7

b

2 a and b

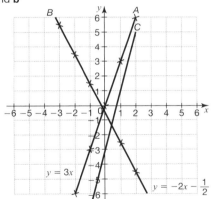

c $y = 4x - 3$

3 a B and C, because they have the same gradient of 2.

b A and B, because they both have a y-intercept at (0, 1).

4 $y = 2x - 4$

5 a 2 **b** 1

6 a 2.5 **b** 4 **c** 0.6

7 a

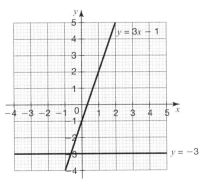

b Any answer between −0.6 and −0.7.

8

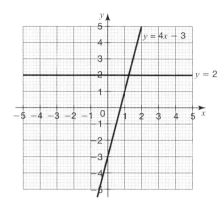

Any answer between 1.2 and 1.3.

Solving simultaneous equations

1 a $x = 3, y = 6$

b $x = 4, y = 1$

c $x = 2, y = -2$

2 a $x = 2, y = 5$

b $x = -1, y = -3$

c $x = 1, y = 6$

3 a $x + y = 2$ $2x - y = 1$

x	0	2
y	2	0

x	0	$\frac{1}{2}$
y	−1	0

b

c

Quadratic graphs

1 a C and D **c** A **e** B

b E **d** D

2 a

x	−3	−2	−1	0	1	2	3
y	6	1	−2	−3	−2	1	6

b

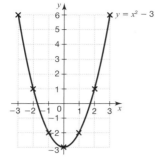

c $x = 0$ **d** $(0, -3)$

3 a

b $x = -1$ **c** $(-1, -2)$

4 a

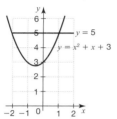

b $x = -2$ or $x = 1$

5 $x = -3$ or $x = 1$

6 a

b $x = -2$ or $x = -1$

c Any answer close to $x = -2.6$ or $x = -0.38$.

Solving quadratic equations

1 a 0 or -6 **b** 0 or 11 **c** 0 or 3
2 a 4 or -4 **b** 9 or -9 **c** 10 or -10
3 a -2 or -3 **b** -5 or 2 **c** 2 or 7
4 a 0 or 3 **b** -5 or 5 **c** -6 or 3

Cubic and reciprocal graphs

1 a A, C and D **b** B **c** E **d** D
2 a

x	-2	-1	0	1	2
y	-4	3	4	5	12

b

3 a cubic **b** $(0, -8)$ **c** $(2, 0)$
4 a

x	-3	-2	-1	$-\frac{1}{2}$	$\frac{1}{2}$	1	2	3
y	$\frac{2}{3}$	$\frac{1}{2}$	0	-1	3	2	$1\frac{1}{2}$	$1\frac{1}{3}$

b

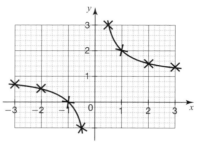

Drawing and interpreting real-life graphs

1 a \$3
b \$2.50
c

d 6 miles
2 a 16 m **c** 8 seconds
b 4 seconds **d** 2 seconds and 6 seconds
3 a 10 m/s
b The cyclist is travelling at a constant speed of 10 m/s.
c $-\frac{1}{2}$ m/s²
d The cyclist stops.
4 a A2; B4; C3; D1
b

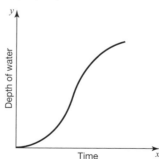

Ratio, proportion and rates of change

Units of measure

1 a 400 cm **b** 5 kg **c** 1500 ml **d** 8.25 km
2 2500 ml
3 a Luke: 240 seconds; Adam: $3 \times 60 + 47 = 227$ seconds. Adam arrived first.
or Luke: $240 \div 60 = 4$ minutes; Adam: 3 minutes 47 seconds. Adam arrived first.
b 13 seconds
4 Ben $= 1.25$ m $= 3.2 + 0.8$ feet $= 4$ feet. Tom is taller.
Or Tom $= 4.8$ feet $= 3.2 + 1.6$ feet $= 1 + 0.5$ metres $= 1.5$ metres. Tom is taller.

Ratio

1 3 : 2   2 £10   3 £30

4

blue	blue	blue	yellow	yellow	yellow	yellow	yellow	yellow	yellow

Phil needs 3 × 500 ml = 1500 ml = 1.5 litres of blue paint.
He has 2 litres of blue paint.
Phil needs 7 × 500 ml = 3500 ml = 3.5 litres of yellow paint.
He has 3 litres of yellow paint.
Phil has enough blue paint, but does not have enough yellow paint.

Scale diagrams and maps

1 10 000 cm; 100 m
2 3 cm
3 20 m
4 24 cm

Fractions, percentages and proportion

1 $\frac{3}{4}$
2 a 1 : 3 : 6
   b $\frac{3}{10}$
   c 60%
3 $\frac{1}{80}$
4 32%

Direct proportion

1 a £16
   b £144
2 a £1.20
   b 2p
   c It is a straight-line graph; the graph passes through the origin (0, 0).
3 a 7 teaspoons of turmeric; 14 teaspoons of chilli powder; $17\frac{1}{2}$ teaspoons of cumin
   b Sally has 75 g of chilli powder. That is 75 ÷ 3 = 25 teaspoons.
     Sally needs 14 teaspoons to make the curry for her class. She does have enough.

Inverse proportion

1 a £400
   b 9
   c £2000
   d Prize money and number of winners are in inverse proportion.
2 a 6 hours
   b 30 minutes (or 0.5 hours)
3 a y decreases (e.g. when $x = 1$, $y = 1$; when $x = 8$, $y = 1\frac{1}{8}$)
   b Inversely proportional, because as x increases, y decreases.
4 10 minutes

Working with percentages

1 40%
2 a 60%
   b 28 (£ millions)
3 a £280.90
   b £281.22

Compound units

1 50 m/s
2 4 minutes
3 6000 Newtons/m²
4 On Saturday Sami drove 4 × 50 = 200 miles; on Sunday Sami drove 356 ÷ 8 × 5 = 222.5 miles. Sami drives further on Sunday.

Geometry and measures

Measuring and drawing angles

1 a 123°
   b 42°
   c 331°

2

3 a 100°, 120°, 140°, 160°
   b first angle + second angle = 87°. This means both angles are less than 87°, and so they both must be acute.

Using the properties of angles

1 a Angle ACB = 52° (Angles in a triangle add up to 180°)
     x = 128° (Angles on a straight line add up to 180°)
   b Angle ADC = 86° (Angles in a quadrilateral add up to 360°)
     x = 94° (Angles on a straight line add up to 180°)
2 a Angle BED = 39° (Alternate angles are equal)
     Angle BDE = 39° (Base angles in an isosceles triangle are equal)
     x = 102° (Angles in a triangle add up to 180°)
   b Angle DCF = 98° (Vertically opposite angles are equal)
     x = 98° (Corresponding angles are equal)
3 Angle CFG = 62° (Co-interior angles add up to 180°)
   x = 66° (Angles on a straight line add up to 180°)
4 $x + 40 + 3x + 5x - 40 = 180°$
   $9x = 180°$
   $x = 20°$
   Angle $BAC = x + 40 = 20 + 40 = 60°$
   Angle $ACB = 3x = 3 × 20 = 60°$
   Angle $ABC = 5x - 40 = 5 × 20 - 40 = 60°$
   Triangle ABC has equal angles of 60°. Therefore, it is an equilateral triangle.

Using the properties of polygons

1 a 720°
   b 120°
   c 60°
2 a octagon
   b all angles are equal; all sides are equal
   c 135°
3 exterior angle = 180° − 144° = 36°
   number of sides = 360° ÷ 36° = 10
   therefore it is a decagon.

Using bearings

1 a 065°
   b 318°
   c 245°
2 096°
3 a 060° (any value 058° to 062° accepted)
   b

Properties of 2D shapes

1 a

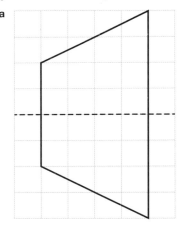

b trapezium
c one pair of parallel sides

2 a

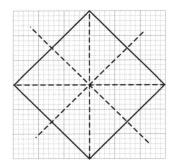

b 4
c square
d Two from: all sides equal in length; all angles are 90°;
diagonals are equal; diagonals bisect each other at 90°

3 a rectangle, rhombus
b

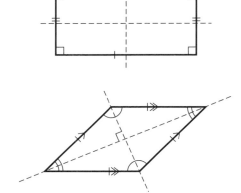

Congruent shapes

1 D and F
2 63°
3 **a** SAS
  b ASA
4 No, they are not congruent. They have the same angles, but
  the sides may not be the same size (one triangle could be an
  enlargement of the other).

Constructions

1 a and **b**

2 a and **b**

3 a

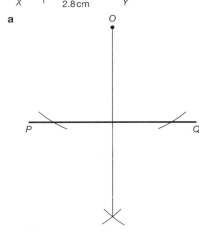

b 2.5 m

Drawing circles and parts of circles

1 a–d

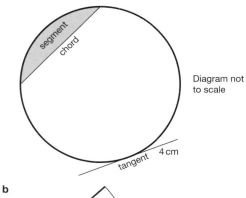

Diagram not to scale

2 a and **b**

Diagram not to scale

3 No, Donald is not correct. A segment of a circle is the area
  enclosed by a chord and an arc; a sector of a circle is the area
  enclosed by two radii and the arc between them.

Loci

1 a

b

c

2

3

4

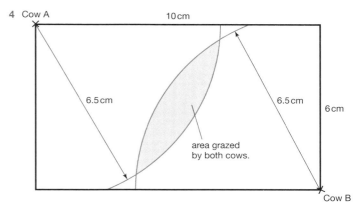

Scale 1 cm : 2 m

Perimeter

1 54 cm
2 11.4 cm
3 Perimeter of cushion $= \frac{1}{2} \times 2 \times \pi \times 24 + 30 + 48 + 30$
   $= 183$ cm (to nearest cm)
   $= 1.83$ metres. No, Greta does not have enough lace.

Area

1 **a** 72 cm²   **b** 22 cm²   **c** 82.5 cm²
2 12 cm²
3 2.63 cm²

Sectors

1 **a** $\frac{1}{3}$ **b** 3π cm²
2 **a** 6.2 cm²   **b** 10.0 cm
3 **a** 8.73 cm²   **b** 3.49 cm

3D shapes

1 a

b triangular prism

c

Number of faces	Number of edges	Number of vertices
5	9	6

2 a

b

c

3 a

b 10

Volume

1   20 cm³
2   **a** 942 cm³   **b** 770 cm³
3   1072.33 m³
4   Volume of tank = 40 × 40 × 60 = 96 000 cm³
Volume of water in tank, 80% full = 0.8 × 96 000
= 76 800 cm³
Height of water in pond (1st fill) = 76 800 ÷ (80 × 60) = 16 cm
Height of water in pond (2nd fill) = 16 × 2 = 32 cm
Height of water in pond (3rd fill) = 16 × 3 = 48 cm
Three tanks of water are needed to fill the pond.
Alternative method: divide volume of pond by volume of water in tank.
$\frac{80 \times 60 \times 48}{76\,800} = 3$

Surface area

1   **a** 6   **b** 136 cm²
2   96 cm²
3   **a** 2463.01 cm²   **b** 301.59 cm²
4   386.42 cm²

Using Pythagoras' theorem

1   $x = 5$ cm; $y = 9$ cm
2   17.9 m²
3   $2\sqrt{5}$
4   diagonal of doorway² = 70² + 190²
= $\sqrt{41\,000}$ = 202.48 cm = 2.0248 m = 2.02 m (2 d.p.)
Yes, the artwork will fit through the diagonal of the doorway.

Trigonometry

1   31.6°   2   22.9 cm   3   4.53 m

Exact trigonometric values

1   **a** 1   **b** 45°
2   2 cm
3   10 cm
4   45° and 45°

Transformations

1

2

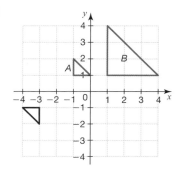

3   Rotation 90° clockwise about (1, −1); or rotation 270° anticlockwise about (1, −1).

4   **a** and **b**

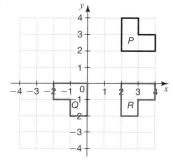

c   Reflection in $y = 1$

Similar shapes

1   **a** *YZW*   **b** 3   **c** 12 cm
2   **a** 37.5°   **c** isosceles
      b 4 cm   **d** 2.5 cm
3   **a** $\frac{2}{3}$ **b** 3
4   **a** $\frac{1}{2}$ **b** 5.6 cm   **c** 46°

Vectors

1   **a** = $\begin{pmatrix} 3 \\ 2 \end{pmatrix}$ **b** = $\begin{pmatrix} -3 \\ 3 \end{pmatrix}$ **c** = $\begin{pmatrix} 2 \\ -4 \end{pmatrix}$ **d** = $\begin{pmatrix} -4 \\ -2 \end{pmatrix}$

2   **a**

b   −**p** = $\begin{pmatrix} 6 \\ -1 \end{pmatrix}$

c   2**p** = $\begin{pmatrix} -12 \\ 2 \end{pmatrix}$

d   2**p** + **p** = $\begin{pmatrix} -12 \\ 2 \end{pmatrix} + \begin{pmatrix} -6 \\ 1 \end{pmatrix} = \begin{pmatrix} -18 \\ 3 \end{pmatrix}$

3**p** = 3 × $\begin{pmatrix} -6 \\ 1 \end{pmatrix} = \begin{pmatrix} -18 \\ 3 \end{pmatrix}$

Therefore, **2p** + **p** = **3p**

3   **a** $\begin{pmatrix} -2 \\ 5 \end{pmatrix}$

b $\begin{pmatrix} 8 \\ 3 \end{pmatrix}$

c

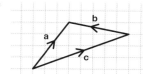

Probability

Basic probability

1   $\frac{1}{10}$ or 0.1 or 10%
2   0.4
3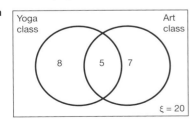
4   $2p - 0.1 + 2p + 0.1 + p = 1$
    $$5p = 1$$
    $$p = 0.2$$

Outcome	Red	Blue	Green
Probability	$2p - 0.1$ $= 2 \times 0.2 - 0.1$ $= 0.3$	$2p + 0.1$ $= 2 \times 0.2 + 0.1$ $= 0.5$	$p = 0.2$

Blue is most likely.

Two-way tables and sample space diagrams

1

	Single	Double	King	Totals
Oak	2	16	12	30
Pine	23	14	17	54
Walnut	1	12	3	16
Totals	26	42	32	100

2   a

		Spinner			
		1	2	3	4
Coin	Heads	1, H	2, H	3, H	4, H
	Tails	1, T	2, T	3, T	4, T

  b $\frac{1}{8}$
  c $\frac{3}{8}$

3   a

	Study sciences	Do not study sciences	Totals
Boys	24	21	45
Girls	35	40	75
Totals	59	61	120

  b $\frac{21}{120}$
  c $\frac{7}{15}$

Sets and Venn diagrams

1   a $\xi = \{21, 22, 23, 24, 25, 26, 27, 28, 29\}$
    b $A = \{21, 24, 27\}$
    c $B = \{24, 28\}$
    d $A \cup B = \{21, 24, 27, 28\}$
    e $A \cap B = \{24\}$

2   a

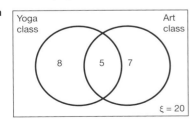

  b $\frac{7}{20}$
  c $\frac{3}{4}$

3   a $\frac{5}{12}$ b $\frac{2}{5}$ c $\frac{3}{20}$ d $\frac{1}{6}$

Frequency trees and tree diagrams

1   a

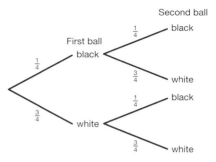

  b $P(W, B) = \frac{3}{16}$
  c $P(B, B) = \frac{1}{16}$

2   a

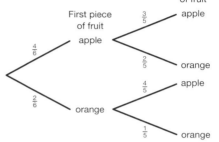

  b $P(O, O)$ or $P(A, A) = \frac{7}{15}$

3   a

  b $\frac{1}{15}$

4   a

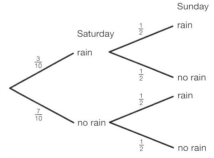

  b $\frac{3}{20}$
  c   50%

Expected outcomes and experimental probability

1   a   50   b $\frac{13}{50}$ c $\frac{3}{10}$ d   20
2   15
3   a $\frac{1}{3}$ b   150   c   45   d   225

Statistics

Data and sampling

1. 65, because that is 10% of 650 (the entire population).

2. $\frac{50}{25\,000} \times 100 = 0.2\%$. The sample is not big enough.

   People in the town centre may not be the only ones using buses. For example, some people may take buses to the local train station, school or hospital.

3. **a** 80 whole cakes

   b Assumptions

   If answered 80, then assuming each person will eat a whole cake; it could affect the answer if instead assume each cake is divided into 8 slices (e.g. then only 10 carrot cakes would be needed).

   Assumed the sample is representative of the population; this could affect the answer because not all the 400 people who have accepted the invitation may turn up.

Frequency tables

1. **a** 30   **b** 91

2.

Number of electronic devices	Tally	Frequency
0–1	|||	3
2–3	ⵏⵏ ⵏⵏ	10
4–5	ⵏⵏ	5
6–7	ⵏⵏ	5
8–9			1

3. **a** Continuous

   b

Mass, m (kg)	Tally	Frequency
$50 \le m < 60$	|||	3
$60 \le m < 70$	ⵏⵏ	5
$70 \le m < 80$	||||	4
$80 \le m < 90$	ⵏⵏ	5
$90 \le m < 100$	|||	3

4. There were 12 people who spent 10 or more minutes, but fewer than 14 minutes, with the GP. Though, none of them may have spent exactly 10 minutes.

Bar charts and pictograms

1. **a** 5   **b** 41

2.

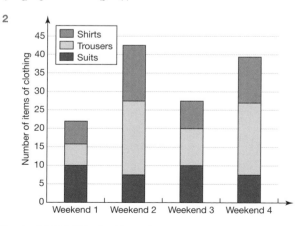

3. **a** 14 species   **b** 2 species

Pie charts

1. **a** 125   **b** 225

2.

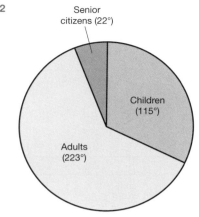

3. 60

Measures of central tendency: mode

1. 3 minutes and 4 minutes
2. $12 < a \le 13$
3. 3
4. Sport (boys); crafts (girls)

Measures of central tendency: median

1. $14\frac{1}{2}$
2. $12 < a \le 13$

Measures of central tendency: mean

1. 11
2. 3 bedrooms (rounded to the nearest whole number)
3. 1 holiday (rounded to the nearest whole number)
4. 46 years old (rounded to the nearest year)

Range

1. **a** 2 years   **b** 3 years
2. **a** 9°C   **b** 15°C
3. **a** Business A: range = £22 255; mean = £34 768

   b Business B: range = £45 354; mean = £38 572.50

   c Either: Business A because its range in profit is lower and the profit is increasing each year, and so it shows a more consistent performance. OR: Business B because its mean profit is higher, and its most recent profit (in Year 4) is £17 432 more than Business A.

Comparing data using measures of central tendency and range

1. **a** 34.5 minutes

   b 28.5 minutes

   c The train journey

   d Either: The bus is better because although it takes longer (on average), the range is lower, and so you can predict the time it takes for the journey.

   Or: The train is better because it is quicker than the bus (on average), although the range suggests it may be less reliable.

2. **a** Mean = 13. This does not represent the age of the people using the playground. In fact, the ages of those using the playground are small children (under 10) and their parents (over 25).

   b There are five modes (3, 4, 5, 7, 8), and so the mode does not represent the age of the people using the playground.

   c Median = 7

3. Mode = 0; Median = 0; Mean = 2 days. Mode or median are the best averages to use, because the mean is skewed by the student who is absent due to sickness for 24 days.

Time series graphs

1 **a** 190 g   **b** 5 days   **c** 20 g

2 **a**

Number of campers on a campsite from March to August

b There is an increase in campers in May. This may be due to May bank holidays, or May half-term, or perhaps there was some very sunny weather.

3 **a**

Sales figures for two different branches of a DIY store

b Branch 1 had a steady increase in sales for the first four months. Then sales levelled off to stay at around £25 000. Branch 2 had a slow start to its sales in the first three months. Then perhaps it had a promotion, because sales increased a lot in month 4. Sales have been increasing ever since.

Scatter graphs

1 **a** Positive correlation. This means as the temperature rises, more pairs of flip flops are sold.

   b Negative correlation. This means as the temperature rises, fewer wellington boots are sold.

2 **a**

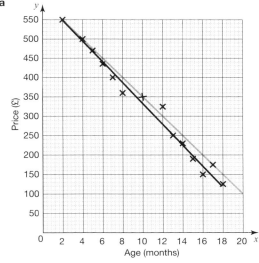

   b The scatter diagram shows a positive correlation between students' maths and physics test percentages. Therefore, the students who got a low percentage in the maths test got the lower percentages in the physics test; the students who got a high percentage in the maths test got the higher percentages in the physics test.

   c The outlier is the point marked at (0, 58)

   d The student was absent for the maths test.

3 **a**

The black line shows the line of best fit.
The grey line shows the line where a laptop loses £150 every 6 months.
The shop owner is not correct. The line of best fit shows on average a laptop loses approximately £159/£160 every 6 months.

   b The line of best fit cannot make a prediction outside the available data. The data only goes as far as 18 months.

Practice paper

Non-calculator

1   3.4
2   C
3   $\frac{10}{14}$
4   $7x - 2$
5   5
6   8%
7   Its opposite sides are equal and it is a parallelogram.
8   900
9   $\frac{7}{12}$
10 **a** 12 and 18
    b 2, 3
    c No. A prime number has exactly two factors: 1 and itself. A multiple of 6 would have at least four factors: 1, 2, 3, 6.
11 5p 5p 10p 50p £1
12 $\frac{5}{12}$
13 $\frac{2a}{b} = \frac{2 \times 6}{4} = \frac{12}{4} = 3$, which is an odd number answer.
14 9.5 m
15 **a** 24   **b** 5 : 6
16 10
17 400
18 £2

19 a

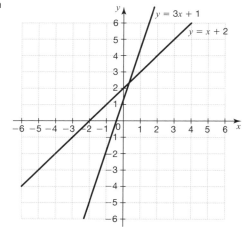

b $x = 0.5$; $y = 2.5$

20 4 kilometres

21 No labels for phone call or letter. Phone call and letter should not be equal sizes – phone call should be twice as big (40°) than letter (20°).

22 a i $-2, -1, 0, 1, 2$ ii $-4, -3, -2, -1, 0, 1, 2$

b 8

23 4.5×10^{-4}

24 a 40% $\left(\text{Or } 0.4 \text{ or } \frac{2}{5}\right)$

b Library A has 28% of 1500 = 420 local history books; Library B has 16% of 3200 = 512 local history books. Therefore, library B has more local history books.

25 $5\frac{5}{8}$

26 a $x = 5$ **b** $3x(x + 9)$

27 $24\,\text{cm}^2$

28 a 7th

b $7n + 1 = 36$

$7n = 35$

$n = 5$

This sequence includes 36 because n is an integer.

29 a 1.43   **b** 0.819

30 $8\pi + 15\,\text{cm}^2$

Practice paper

Calculator

1 11

2 A and C

3 -3.2

4 £11.50 × 37 = £425.50. B £462.50 is the better rate of pay.

5 1.48pm and 35 seconds

6 $(4, -1)$

7 a 15

b 17.5

c Median because it better represents the 'typical' score – the mean is skewed by the low score of 2.

8 0.8

9 a

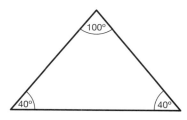

b isosceles

10 50 cm

11 a £34 000

b Profit for Branch 2 = £90 000 − £52 000 = £38 000. Profit for Branch 3 = £44 000 − £20 000 = £24 000.

The manager is incorrect. He has not made double the profit.

12 a $x = 100$ **b** $3a^2 + 2ab$ **c** $c = \frac{p}{t}$

13 £210

14 12

15 a $x = 34°$; $y = 68°$.   **b** kite

16 £36

17 a Positive correlation

b Student should draw a line of best fit and read off the number of deckchairs at 20°C. An answer between 42 and 46 is acceptable.

18 a

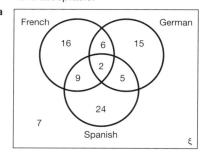

b 84

c $\frac{1}{12}$

19 £450

20 $146.61\,\text{m}^3$

21 $\begin{pmatrix} -2 \\ 6 \end{pmatrix}$

22 a £393.08   **b** £57.60

23 9.42 cm

24 x is a negative number. Reason: x^3 is a value of x multiplied by itself three times.

negative × negative × negative = negative.

25 a

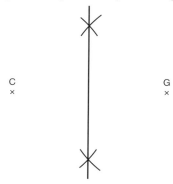

b 0.82 km (0.8 to 1 d.p.)

26 a 26.287   **b** $5^2 + \sqrt[3]{8} = 25 + 2 = 27$

27 £9373

28 a $y = -2x + 2$ **b** $y = -2x$